JOURNEY'S END

LOCOMOTIVES AND ROLLING STOCK FOR SCRAP IN GREAT BRITAIN

GORDON EDGAR

AMBERLEY

Right: Condemned locos at Stockport Edgeley motive power depot (MPD) in May 1963, with Stanier 'Jubilee' 45678 *De Robek*, Hughes/Fowler 'Crab' 42921, and Fowler '4MT' 2-6-4 tanks 42416, 42357, and 42372 pending their final journeys to the breaker's yards. (Author's collection)

Front cover: Class '40' 40108 was the first of many diesel locos to be gathered at Swindon works for so-called 'long-term storage' in the early 1980s. Taken out of service in August 1980, it was reduced to scrap metal during the following December, just two months after being presented on display at the 5 October 1980 works open day, in the company of D1015 *Western Champion*, which happily survived. (Author)

Back cover: Row upon row of locos dumped at Dai Woodham's scrapyard at Barry on 15 September 1968. Remarkably, most of these never saw the cutter's torch and many have since enjoyed further active service. (Author)

First published 2024

Amberley Publishing
The Hill, Stroud
Gloucestershire, GL5 4EP

www.amberley-books.com

ISBN 978 1 3981 1847 8 (paperback)
ISBN 978 1 3981 1848 5 (ebook)

British Library Cataloguing in Publication Data.
A catalogue record for this book is available from the British Library.

Typeset in 9.5pt on 11.5pt Sabon LT Std.
by SJmagic DESIGN SERVICES, India.
Printed in Great Britain.

CONTENTS

INTRODUCTION & ACKNOWLEDGEMENTS

As a convicted lifelong railway enthusiast, possessing an avid devotion stretching back to some of my earliest childhood recollections of over six decades ago, I know that I also share with many others a fascination for both photographing and whenever possible travelling on timetabled services or charter trains entrusted to locomotives facing the end of their working lives; or perhaps even taking a journey in a multiple unit fast approaching the end of its revenue-earning service, as that inevitable fateful date of withdrawal from service draws ever closer. Such a curiosity extends further, moreover: interesting photographic opportunities were often to be exploited when visiting depots and yards, seeking out condemned locomotives and rolling stock facing that final journey to a breaker's yard, or perhaps even photographing an arrival at a predetermined place of disposal, and occasionally with luck permitting witnessing the 'final rites' taking place. Some locomotives, multiple units, and rolling stock could be facing prolonged storage, pending cannibalisation for reusable components before their final onward despatch, and others might simply be hastily cut up by contractors on site just where they stand, deemed as being unsuitable to make just one final main line journey to a scrapyard. All such facets around the 'journey's end' of locomotives and rolling stock have offered compelling photographic potential over the years to so many railway enthusiasts. Capturing the final days of regular use, preceding withdrawal from service, subsequent storage, and eventual disposal of redundant, condemned, or quite simply worn-out and woebegone railway locomotives, steam, diesel, and electric traction alike, has long proven to be an absorbing photographic pursuit for the enthusiast, and is arguably an important associated facet of future railway history. It is therefore deserving of documentation and closer attention. This book is predominantly a photographic portfolio, but with informative captions setting the scene, broadly spanning the timeline between the last months of steam operations in the various regions of the United Kingdom, through to the disposal of some of those very locomotives and multiple units that ousted main line steam traction under BR's Modernisation Plan, following the recommendations of the so-called British Railways Board report *The Reshaping of British Railways*, released in 1963.

The photographs and captions in this book are presented in a similar fashion to those that have featured in the previously published ten regional *Industrial Locomotives & Railways* titles, a series by the same publisher in which I have examined the industrial railway scene around Great Britain, essentially during the last half a century. The opening chapter reviews the last months primarily of steam traction around all regions of the United Kingdom, with poignant scenes of woebegone locomotives in their final days, or those having already been cast aside in the name of progress and awaiting that inevitable fate. The famous Barry scrapyard of Dai Woodham, which unlike many of the other one hundred and more BR-approved contractors authorised to dispose of its locomotives, was easily accessible, and therefore receives the attention that it unquestionably deserves, covered in its own chapter. This is especially appropriate in the light of Dai Woodham himself having played a unique and already well-documented part over several decades in orchestrating the saving of over two hundred former BR steam locomotives, many to be found today on Britain's heritage railways, most of those having since been restored to full working order and presented in their deserved splendour once again.

Traditionally, BR had scrapped redundant locomotives and some rolling stock at its own workshops, but the sheer number requiring disposal during the 1960s and thereafter led to sub-contracting this work out to numerous authorised private commercial breakers. These yards varied in the scale of their operations, but Dai Woodham at Barry, Cashmore and Buttigieg at Newport, Cohen at Cransley, and Draper at Hull, although not exclusively, were high on the list of major players in this field during the closing years of BR steam traction disposal, with Booths of Rotherham, A. King & Sons of Norwich and Snailwell, and Vic Berry of Leicester, being the principal contractors handling locos and vehicles during the following decades, the latter two being accredited specialists in blue asbestos removal and disposal from the many BR multiple units and coaching stock built during the 1950s and 1960s era and containing this hazardous contaminating material.

I have endeavoured to provide as much accompanying caption detail as practicable within the scope of a book this size and format. Sadly, space will not allow coverage of all aspects of this subject, however a representative selection of photographs has been carefully chosen to convey as balanced a picture of this subject as is practicable, with all BR main line non-steam traction featured herein categorised as 'heritage' in nature, predominantly of 1950s or 1960s design and build, and in some cases older, but two appropriate exceptions are also covered, the ambitious prototype Advanced Passenger Train (APT-P) upon which today's tilting train technology was based, and the prototype of the outstandingly successful and enigmatic High Speed Train (HST), with some still in service during 2024. I am indebted to Roy Burt, Charlie Cross, and Kevin Lane for contributing relevant photographs from their collections, without whose assistance this book would not have been possible. I have also benefitted from the assistance of the Railway Correspondence & Travel Society and the Industrial Railway Society, and the following invaluable sources have been consulted during the compilation of this book:

Veteran English Electric Vulcan Foundry 1965-built BR 'AL6' Class E3192, latterly Europhoenix Class '86' demonstrator 86247, languishing at Freightliner Crewe Basford Hall depot in the rain on 28 November 2015, having been used as a spares donor to Freightliner 86501 standing behind. It was cut up there during the following week, but both cab ends survived. Several Class '86' locos were acquired by Europhoenix for conversion and resale to Bulgaria, although 86247 only received lighting and front-end modifications and was never reactivated by the company, acting purely as a demonstration loco, presented in its distinctive corporate livery. (Author)

BR Steam Locomotives Complete Allocation History 1948–1968 by Hugh Longworth

BR database – the complete British Locomotive database 1923–1997

Blood and Custard – a railwayman's website for railwaymen.

Gordon Edgar
Ripon, North Yorkshire

1

AN INEVITABLE FATE

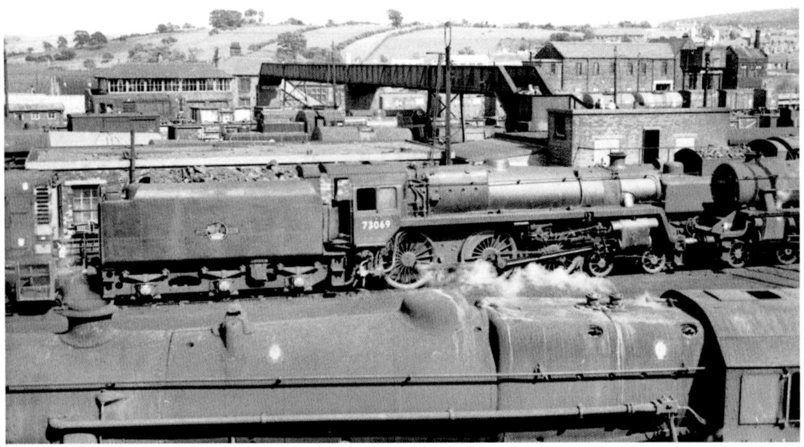

Above left: Two rows of BR Standard '4MT' 4-6-0s at Carnforth MPD on a damp 30 December 1967, cast aside and awaiting their final journeys to the scrapyards. To the left of the Ford Anglia is 75058, withdrawn from service in December 1967, languishing along with other class members in this sombre-looking line-up at the depot's north end sidings. All were Skipton-allocated between March 1965 and April 1967, moving to Carnforth for their final months of BR service. 75058 was despatched within days of this photo having been taken to the by then just closed Carlisle Kingmoor steam depot en route for scrap at Arnott Young's yard at Carmyle, where it was hastily disposed of during April 1968. (Author's collection)

Below left: Carnforth MPD in May 1968, with what was the last BR Standard Class '5MT' 4-6-0 73069 remaining in traffic, and surviving until the end of BR steam. In clean external condition, it was sought-after motive power on some of the farewell to BR steam railtours in 1968, but surprisingly it was not saved for preservation. It was seen in the company of two active Stanier 'Black Fives' and the preserved LMS maroon-liveried Ivatt '2MT' 2-6-0 6441 can just be glimpsed beneath the Warton Road footbridge. (Author's collection)

Opposite page: The closing chapter of the BR steam era was unfolding as locos congregated at Carnforth depot at the end of the last full year of steam, on an appropriately cold, damp and murky Saturday 30 December 1967. Centre-stage was 'Britannia' Pacific 70021 (formerly bearing the name *Morning Star*), in the company of Standard '4MT' 75015 to its front, and four BR Standard '9F' Class 2-10-0s, 92110 and 92223 being identifiable. 70021 and 92110 had just been despatched under their own power from Carlisle Kingmoor MPD, which was closed to steam that weekend. Both, along with others from Kingmoor, were immediately officially withdrawn from service upon arrival. Carnforth depot would never witness such a number and variety of BR Standard Class locos again, and in just a little over five months, as the attrition was to set in for the remaining active steam locos over the ensuing months, the end finally came in August 1968. (Author's collection)

70021

BR 'Metro-Vick' Class '28' D5712, heading an up empty block oil train, rounds the curve at Furness & Midland Junction, Carnforth, in July 1968, just weeks before all remaining class members were to be taken out of service en masse and placed into store. The local railwaymen amusingly often referred to the 'Metro-Vicks' as 'Wonder Loaves', the brand name of a popular consumer food product at this time. (Author's collection)

Left: An overall view from Carnforth's Warton Road footbridge, over which many railwaymen and railway enthusiasts alike had crossed over the years. English Electric Type 4 and Sulzer Type 2 locos had already infiltrated the depot in this scene on Saturday 30 December 1967. In the foreground were a trio of BR Standard Class '4MT' 4-6-0s, 75021, 75009 and 75034, whilst beyond them is Class '40' D218 and another class member. Steam locos simmer in the shed yard beyond, some having arrived from the then just closed Carlisle Kingmoor steam MPD, but most would never again move a wheel under their own steam. (Author's collection)

Right: On Saturday 18 November 1967, a Stanier '8F' 2-8-0 rests outside the southern end of Carnforth MPD, as another 'shed mate' heads off into the mist, passing its iconic concrete coaling tower, now the last such survivor in the UK, although today in poor condition, facing an uncertain future, just as the recipients of its contents had done so over half a century before. (Eric F. Bentley – author's collection)

45017

Above: Lostock Hall MPD on a cold and misty winter's morning in January 1968, with many Stanier '8F' 2-8-0s and 'Black Five' 4-6-0s present, realistically facing little prospect of further work. 48492 was to be officially withdrawn from service during the following month and following a further two months of storage at Lostock Hall depot was towed off to Draper's Hull scrapyard, then cut up during May 1968. (Author's collection)

Opposite page: On Saturday 27 July 1968, the end was nigh for BR steam, but the replacement 'Class '17s' were also facing minimal service in the future. With a full tender of coal, Stanier 'Black Five' 4-6-0 45017 was moving away from Carnforth depot's coaling tower, passing the doomed 'Claytons', BR blue D8507 and two-tone green D8506, plus two 350hp diesel-electric shunters. Following migration to Scotland during 1969, both 'Claytons' were soon to be taken out of service, during 1971. (Author's collection)

Above: Lostock Hall MPD yard in early 1968, during the depot's final year and the fast-approaching end of BR steam traction, with lines of condemned locos awaiting despatch to the breaker's yards.

Opposite page: Rose Grove MPD in August 1968, during the week following cessation of BR steam activity and the depot's closure. Stanier 'Black Fives' 44690 and 44899 were cold and unceremoniously dumped alongside the running shed building, awaiting despatch for scrapping, at Ward's at Beighton and Cohen's of Kettering, respectively. The simple inscription chalked across 44899's grimy tender side kept the sentiment quite plain and simple: 'Steam Forever'! (Both author's collection)

44690
44899

Left: A regular turn for a Lostock Hall-allocated Stanier '5MT' 4-6-0 during the final months of BR steam in early 1968 was the 16:53 (Sundays only) Preston to Liverpool Exchange passenger diagram, including through coaches from Glasgow and Edinburgh in its formation. Just prior to departure time, in late April 1968, 'Black Five' 45212 had a good head of steam, ready for departure, as a conversation between enthusiastic railfans, some about to experience one of their last BR steam-hauled journeys, was taking place on the platform alongside. Happily, 45212 is still with us today as a main line performer.

Opposite page: With merely a handful of weeks remaining before its withdrawal, Stanier '8F' 48392 heads a mixed freight past Bolton MPD in April 1968. Reallocated from Lostock Hall during the previous month, it almost saw out the end of steam at Bolton depot, withdrawn in the penultimate week before its final closure, on 30 June. The Horwich 1945-built heavy freight loco was cut up at Draper's of Hull in the September. (Both author's collection)

Right: Glimpsed through the broken windows of an abandoned bothy at Heaton Mersey MPD on 24 April 1968, one week before its closure, '8F' 2-8-0s 48344 and 48197 repose, both forsaken and bereft of smokebox number and shed plates, which quite possibly by that time were already in the hands of avid collectors of Railwayana. A one-way journey was in prospect for these 'Eight Freights', 48344 despatched to Buttigieg's at Newport, and 48197 to Draper's at Hull, where they were hastily broken up, in the July and August, respectively, of that same year. (Author's collection)

48392

48773
1L43

Opposite page: The 'Farewell to BR Steam' 1L43 railtour, photographed upon its return to Manchester Victoria on 28 July 1968, the now-preserved Stanier '8F' 48773 having worked the final leg of the chartered train from Rose Grove, via Copy Pit, Bury and Bolton. The station pilot, Sulzer Type 2 D5209, can be glimpsed between the awning supports. The relaxed, 'turn a blind eye' approach to the railfans' antics during that era is exemplified in this quite remarkable and reckless scene by today's standards, which unfolded at this principal British main line station! (Author's collection)

Right: The end of steam traction was fast approaching in this scene at Stockport Edgeley MPD on 28 October 1967. A depressing-looking collection of withdrawn Stanier locos were lined up, with 'Black Five' 4-6-0 45346 at the head, awaiting despatch to the breaker's yards, as an active Stanier '8F' 2-8-0 clanked by in charge of a block cement train, its working days also very much numbered. (Eric F. Bentley – author's collection)

Left: By February 1968, former Crosti-boilered Class '9F' 2-10-0 92024 had gravitated to the loco dump at Speke Junction depot yard, where many locos taken out of service from various depots in the North West had been assembled, prior to disposal to their allotted scrap merchants. The former Birkenhead BR Standard ten-coupled workhorse stood in the company of 'Britannia' Pacific 70024 (formerly named *Vulcan*), latterly of Carlisle Kingmoor MPD. Both had been taken out of service in late 1967. Within three months, both had completed their final journeys, 92024 to Campbell's yard at Airdrie, and 70024 to Ward's at Killamarsh. (Author's collection)

Above left: BR Standard Class '2MT' 2-6-0 78062 stood condemned at Bolton MPD in July 1967, having seen less than eleven years in BR service. Withdrawn during May 1967, it was to remain at Bolton until September, when it was finally towed away for disposal at Draper's of Hull, and was rapidly dealt with during the October. The smallest of the BR Standard classes, the entire batch of sixty-five locos were built at Darlington Works between 1952 and 1956; 78062 was one of the last three to emerge from that works, in October 1956. They were fitted with tender cabs to enhance visibility when running tender-first, also offering some degree of crew protection from the elements. First allocated to Wigan Central (L&Y) depot, 78062 was employed on local stopping passenger services on the Rochdale, Bolton, Southport, and Liverpool lines. Four class members were despatched to Woodham's Barry scrapyard, with 78018, 78019, and 78022 having since been restored to working order. 78059 is now providing the basis for a 'new build' Standard Class '2MT' 2-6-2 tank, BR's tank version of this class of which none survived, and it will take the next number in the class sequence, 84030. (Charlie Cross)

Below left: Bolton MPD closed to steam on 30 June 1968, and the last Stanier 'Black 5' 4-6-0 and '8F' 2-8-0 locos to remain active at that depot were standing cold, awaiting their final moves for breaking up. 45104 was the last active steam loco to arrive back on shed, having been allocated to the Castleton pilot duties on Saturday 28 June. 48652 had also been out during that day and had returned on shed a couple of hours before the 'Black Five'. Both were despatched for scrap by the September, 48652 being quickly broken apart at Draper's Hull facility during the following month, and 45104 at Cohen's of Cransley, near Kettering, by early 1969. (Author's collection)

Opposite page: A thoroughly wet and bleak day at Northwich MPD, Saturday 13 May 1967, during the last full year of BR steam traction. Workaday Stanier '8F' Class 2-8-0 freight engines 48063, 48334, 48205, among others, fill the air with drifting smoke as they simmer on shed. That unique combination of senses experienced when visiting such run-down steam running sheds during their final months, the air being infused with the aroma of coal smoke, hot oil, and steam, conveying a heightened sense of excitement, but tinged with the sadness at the thought of steam's imminent demise. Northwich shed was officially closed to steam traction on 4 March 1968, but continued in use as a diesel stabling point and for servicing heritage main line steam locos until 1987. It was finally demolished in 1991. (Author's collection)

48334

Above left: Former Beattock banking loco, BR Standard '4MT' 2-6-4 tank 80005 abandoned on the 'scrap line' at Eastfield MPD in September 1966. The BR Derby Works-built loco had served the Scottish Region throughout its entire career, supplied new to Kittybrewster MPD in October 1952, subsequently allocated to Corkerhill, Ayr, Beattock, and finally Polmadie. Withdrawn in August 1966, it was cut up at Motherwell Machinery & Scrap, Wishaw, in October 1966. Type 1 'Clayton' D8505 was standing behind, itself to be withdrawn from service just five years later, eventually being scrapped at BR's St Rollox works during May 1974 after an extended period of storage in the company of several other rejected class members. (Charlie Cross)

Below left: The sun was setting on Gresley 'A4' Pacific 60034 in September 1966 at Perth depot. Once a magnificent resplendent East Coast Main Line (ECML) 'racehorse', but latterly working out of Aberdeen Ferryhill depot, it was destined never to turn a wheel under its own steam again. Having just been withdrawn from service during the previous month, and with its *Lord Faringdon* nameplates and smokebox numberplate already torched off, all that remained for it was a one-way journey to the breaker's yard. This was soon to come about, and it was duly despatched to the Hughes Bolkow scrapyard at North Blyth in the November of that year. It was to be no more by the end of January 1967. Three North British 0-4-0 diesel-hydraulic shunters stood behind, including 1958-built D2729, also destined to be scrapped within one year. Such was the utter waste of BR's motive power resources during the 1960s, inherently due to falling traffic levels and the wholesale closure of many local freight facilities and goods yards countrywide, influenced by the increasing road haulage competition of the time. Alas, this was also influenced by BR having procured various shunting loco types, many deemed to be 'non-standard' long before their life expiry. (Charlie Cross)

Opposite page: A late 1960s photo of Northern Ireland Railways Class 'WT' 2-6-4 tank *No.6*, banked by a sister loco, working a loaded spoil train of twenty hopper wagons alongside Belfast Lough, between Magheramorne cement works and Greencastle, near Belfast. The rock fill traffic ran from early 1966 until May 1970, in connection with the M2 motorway construction. Eighteen 5ft 3in gauge Ivatt 'WT' Class locos were built at Derby Works between 1946 and 1950 for NCC (LMS), based on the LMS Fairburn 2-6-4 tank design. Bestowed the nickname 'Jeeps', their original duties included working passenger services between Belfast, Larne, Portrush, and on the line to Derry via Coleraine, before ending their days on these spoil trains, notably the last regular main line steam-hauled freight trains in the British Isles. (Author's collection)

Above left: A loco that could well be described as illusive and not widely known of was the sole 5ft 3in gauge Harland & Wolff (H&W) Class 'X' 330hp 0-6-0 diesel-hydraulic, numbered '17' and named *Harlandic*. Built in Belfast in 1936, it was one of several pioneering and innovative designs of diesel loco built during the 1930s and 1940s by H&W, and marketed under their 'Harlandic' trademark. It was initially leased to the Northern Counties Committee (NCC) and was used on the Belfast Harbour Commissioners private lines, and on NCC's own sidings in Belfast. It was taken into capital stock in 1941 when purchased by NCC, which subsequently passed into the ownership of the Ulster Transport Authority (UTA), then in 1967 to Northern Ireland Railways (NIR). It is believed to have seen little if indeed no further use when under NIR ownership, and was eventually cut up in mid-1970. It was photographed stabled on shed in Belfast during happier times, either in the late 1950s or early 1960s, when under UTA ownership.

Below left: North Tyneside 600v DC Motor Parcels Van E29468E at Newcastle Central circa 1965. Two such units were built for the LNER by Metropolitan-Cammell in 1938. All electric units on this third rail system had been withdrawn by June 1967 and replaced by diesel multiple units, a rare example of the abandonment of main sections of a previously electrified railway route in Britain, although ironically much of this Tyneside network was to be re-electrified as the Tyne & Wear Metro using 1,550v DC overhead, those lines opening in 1980.

Opposite page: An ideal candidate for a caption competition, and it is probably safe to assume that the driver of this workaday eight-coupled freight loco, 63455 at the dilapidated Tyne Dock MPD in this summer 1965 candid photo, was not at all impressed with the depot fitters! The veteran Armstrong Whitworth-built Raven 'Q6' 0-8-0, new to Neville Hill MPD in December 1920, was eventually taken out of service on 22 June 1967 and despatched to Garnham, Harris & Elton's scrapyard at Chesterfield, in August 1967. It had worked from Consett depot between June 1943 and May 1965, and in all probability had seen its best days by the time it was transferred to Tyne Dock depot for its final two years of service. Just one member of this pre-Grouping 'Q6' Class survived into preservation, 63395 latterly with BR at Sunderland depot, and thanks to the efforts of the North Eastern Locomotive Preservation Group is today a worthy member of the North Yorkshire Moors Railway (NYMR) loco fleet. (All author's collection)

63455

A line-up around one of the turntables at York roundhouse in early 1967, shortly before the depot was closed to steam traction, including 'B1' Class 4-6-0 61337, 'K1' 2-6-0 62001, 'Jubilee' 4-6-0 45675 *Hardy*, and 'B1' 4-6-0 61021 *Reitbok*. The National Railway Museum was established in 1975 on the footprint of No.3 and No.4 sheds of this once extensive depot. (Author's collection)

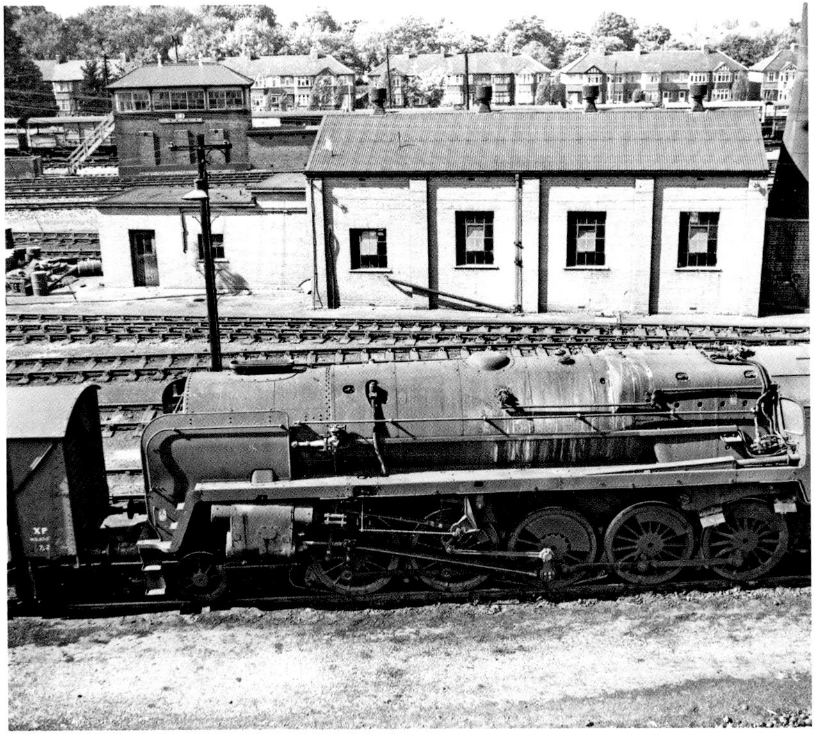

Left: Giesl ejector-fitted Riddles BR Standard '9F' Class 2-10-0 92250 set aside out of traffic at Southall depot in June 1964, just prior to its transfer to South Wales. It was finally withdrawn from Gloucester depot in December 1965, and was cut up at Cashmore's Newport scrapyard in July 1966. Bearing the distinction of having been the last steam loco built at BR Crewe Works, it was delivered new to Banbury depot in December 1958 and was to complete a mere seven years of active BR main line service. (A.E. Durrant – author's collection)

Right: A summer 1964 photo of the loco disposal sidings at Southall MPD, which was used as a holding point for locos from BR's London Division of the Western Region (WR) that were pending despatch for scrap, mainly to BR's approved scrap metal merchants in South Wales, although due to the volume of steam loco disposals demanded at that time, some were also dealt with at Swindon Works. The depot building at Southall has since been retained as a base for some of today's main line steam charter operations centred on the capital and around the south and west of England. (Author's collection)

Worcester-based 'Castle' Class 4-6-0 7011 *Banbury Castle*, by then already missing its smokebox numberplate, but still bearing name and cab-side cast plates, was being turned at Southall depot in early 1964. The 'Castle' was withdrawn from service in the spring of 1965 from Oxley depot, followed by its rapid disposal at Cashmore's yard at Great Bridge, also in the West Midlands, during May 1965. (A.E. Durrant – author's collection)

Collett 'Hall' Class 4-6-0 4920 *Dumbleton Hall* of Bristol Barrow Road MPD gets underway from Eastleigh in charge of the 'Pines Express' inter-regional passenger service in September 1965, which it had worked through from Banbury. Eastleigh's BR Standard '4MT' 2-6-4 tank 80150 was alongside, heading away on the Fareham line. This was 4920's last year of BR service and it had become a celebrity as the longest-serving member of its class, outliving most of its classmates on the main line by at least two years. Following withdrawal from Oxford depot in December 1965, 4920 was despatched to Woodham's Barry scrapyard, but in 1974 was rescued for preservation, recognised as the oldest surviving member of its class. (Author's collection)

Above left: An appropriately murky 20 October 1965 found Collett 'Hall' Class 4-6-0 6917 *Oldlands Hall* in the company of BR Standard '5MT' 4-6-0 73013 and 'Modified Hall' 7912 *Little Linford Hall* at Banbury MPD. Swindon Works 1941-built 6917 had been withdrawn from service at Banbury during the previous month and was cut up on site by J. Friswell Ltd in January 1966. 7912, new to Tyseley depot from Swindon Works in March 1950, had just been taken out of service and was moved to Cashmore's yard at Great Bridge for disposal, being broken up there in January 1966. Remarkably, 73013 made an escape to Bolton depot in April 1966, but it is questionable whether it saw any further work there, officially being withdrawn from service in the May. It then made one final journey to the West Midlands and was also broken up at Cashmore's Great Bridge yard. Due to the extraordinary number of BR locos being taken out of service at WR depots at that time and requiring rapid disposal, the sidings on the west side of Banbury depot were given over for loco disposal purposes by the local private contractor J. Friswell Ltd. Between November 1964 and March 1965, the company dealt with fourteen locos on site, some being deemed unfit for any further movement away from the depot.

Below left: Stanier '8F' 2-8-0 48706 of Bath Green Park MPD coasts into Evercreech Junction station heading the 'Great Western Society Special' from Bath Green Park to Bournemouth via the Somerset & Dorset (S&D) route on 5 March 1966. The former LNER-allocated freight loco was taken out of service at Bath Green Park one week later and was rapidly disposed of at Buttigieg's Newport scrapyard during May 1966. The S&D through route, much lamented but also over time having been derided by some as 'the slow and dirty', was closed two days after this specially chartered train had passed over its hallowed metals. Evercreech station, as it was simply named when first opened in 1862, was situated along the original S&D route between Burnham-on-Sea and Broadstone, 4 miles south-east of Shepton Mallet. It was very soon thereafter renamed Evercreech Junction when, in 1874, the extension north to Bath was opened. All that remains today as evidence of this railway is a business park which now occupies the former goods yard site at this minor Somerset town. The residual main station building has since been given over to private residential housing, and the former track bed has now found use as gardening plots for the residents. (Both author's collection)

Right: The end of Southern Region (SR) steam was nigh in this scene at Nine Elms MPD in 1967, with Bulleid Pacifics shorn of smokebox number and nameplates awaiting an uncertain, but almost inevitable future of disposal at the Newport scrapyards of Messrs Cashmore and Buttigieg. Identifiable is 34021, formerly named *Dartmoor*, which survived until the end of SR steam, working the last steam passenger service from Southampton Docks on 9 July 1967. After reaching Waterloo, it immediately ran light to Salisbury and joined the lines of discarded locos being accumulated at the depot pending scrap, retrieved from the then few remaining SR steam depots. (Author's collection)

Left: Salisbury MPD on 6 August 1967, with BR Standard '4MT' 2-6-4 tank 80146 and 'USA' 0-6-0 tank 30071 awaiting their fate. On 18 June 1967, 80146 had worked the final BR steam-hauled passenger service on the Swanage Branch. Despite its very presentable external condition and having seen less than eleven years of service, it was despatched to South Wales for scrap and disposed of during early 1968 at the Risca yard of Messrs Bird. 30071, latterly a Southampton Docks shunter, was scrapped at Cashmore's Newport yard in March 1968. (Author)

41287
4
3

Opposite page: Ivatt '2MT' 2-6-2 tank 41287 of Guildford MPD departs from Guildford working a branch line service to Horsham, just before its closure. The branch of just over 19 miles in length was known as 'The Cranleigh Line' and was closed on 14 June 1965, ironically four months before its centenary. It was to be the only Surrey railway route to succumb under Beeching's infamous 'Reshaping of British Railways' programme. (Alan Orchard – author's collection)

Above right: A June 1967 glimpse of BR Standard Class '3MT' 2-6-0 77014 languishing inside the dark confines of the seven-track straight shed section of Guildford depot, built into a compact corner of an excavated chalk cliff, adjacent to the appropriately named 845 yards-long Chalk Tunnel, south of the station. This was merely days before its withdrawal from service, simmering in the company of Birmingham Railway Carriage & Wagon Co. (BRCW) Type 3 'Crompton' D6551. A remarkable move, believed to be influenced by two railtours on the SR in April and May 1966, saw 77014 transferred from Northwich depot to Guildford, after which it failed to be returned, thereby enabling it to see out the final year of SR steam. Indeed, in February and April 1967, it also provided motive power for the London and suburbs area legs of two Locomotive Club of Great Britain railtours. 77014 was one of only twenty class members, all built at Swindon works, a class not used on the SR before, their sphere of operations being chiefly confined to the north of England and Scotland. It attained the dubious honour of working the last scheduled steam-hauled service on the SR, a late evening van train from Bournemouth on 9 July 1967, after which it was immediately taken out of service. It was eventually cut up at Bird's Risca scrapyard during December 1967. (Author's collection)

Below right: On the evening of 14 July 1967, a few days after the end of SR steam, the last two steam locos remaining at Feltham depot, by then inactive, were patiently awaiting their removal. Standing in the yard by the disused coaling tower, with coupling rods removed in preparation for their final journey to the scrapyard via Salisbury depot, were BR Standard Class '4MT' 2-6-4 tank 80154 and '4MT' 2-6-0 76058. Brighton Works-built 80154, the final member of its class to be constructed, entered service at Brighton depot in March 1957 and was withdrawn from Nine Elms in April 1967. Doncaster Works-built 76058 entered service in June 1955 at Redhill and was taken out of service at Guildford in March 1967. Both were disposed of by Buttigieg of Newport later that year. (Author)

The eleventh hour of SR steam traction on the LSWR main line through Clapham Junction in June 1967. Just as BR Standard Class '4MT' 80140 was running into the station's carriage sidings with the empty coaching stock of the unadvertised Post Office 'Kenny Belle' passenger service following the morning peak-time shuttle to and from Kensington Olympia, Bulleid 'Battle of Britain' Pacific 34087 *146 Squadron* thundered through the busy station's platform 9, heading a Waterloo-Bournemouth express. The 1956-built 2-6-4 tank was taken out of service at Nine Elms depot on 9 July 1967, moved to Salisbury for despatch to South Wales, and was cut up at Bird's Risca yard in the November of that year. The 1948-built 'Battle of Britain' Class Light Pacific was also withdrawn from service on the last day of steam on the SR, and following short-term storage at Salisbury, it was despatched to Cashmore's Newport yard and disposed of during May 1968. (Author's collection)

Right: Shelf Sidings, alongside Sutton Bridge Junction at Shrewsbury, was a well-used transit point in the 1960s for steam locos making their final journey from the depots in the North West to the South Wales scrapyards for disposal. On 18 August 1968, a line of four Stanier '8F' 2-8-0s were waiting to be dragged on their final leg of the journey to South Wales. In the row from the front were 48450, withdrawn from Edge Hill in September 1967 and previously stored at Speke Junction depot; 48510, withdrawn from Lostock Hall in January 1968 and stored there; 48637, withdrawn from Lostock Hall in September 1967 and initially stored there, and then at Springs Branch; and 48061, withdrawn from Rose Grove in September 1967, initially stored there, and then at Springs Branch. All were destined for Buttigieg's of Newport and were subsequently dealt with at their scrapyard during November 1968. (Author)

Left: Former WR green-liveried, and latterly Patricroft depot's Riddles BR Standard Class '3MT' 2-6-2 tanks 82003, 82031 (and 82034 out of view), residing at Dock Street Goods Yard, Newport, in August 1967, awaiting their call forward to Cashmore's scrapyard for disposal as and when space would become available. Both were withdrawn from service in December 1966, and it was to be twenty-two months before their call finally came. 82003 and 82031 had both spent time allocated to Barry MPD in 1953–54 and, had they returned to Barry for disposal, they would undoubtedly have been with us today, but not one example of this class was purchased by Dai Woodham. Happily, this deficiency is being rectified with a new-build project by 'The 82045 Steam Locomotive Trust', providing a versatile class of loco, and an ideal choice for use on today's heritage railways. (Charlie Cross)

2

THE BARRY SCRAPYARD LEGEND

Today's Heritage Railways' Salvation

The scrap merchant Woodham Brothers had begun life as a company in 1892, dealing in old rope and wood from decommissioned ships in Barry Dock. By the mid-twentieth century, the company was dealing predominantly in scrap metal, recycling the materials for use in the British steel industry. By the late 1950s, the founder's son, Dai Woodham, had seen a lucrative business opportunity in the disposal of some of the increasing number of BRs steam locos and rolling stock being taken out of service and becoming available under its so-called Modernisation Plan of that time. The ease and quick profitability from the scrapping of rolling stock, instead of the more time-consuming task of dismantling locos and the consequent materials separation process, saw the loco stockpile awaiting disposal at Barry progressively increase in number during the 1960s, although some modest inroads of mainly former GWR locos had been made between 1959 and 1965. As the end of BR steam traction in August 1968 became imminent, interest from enthusiasts in the accumulation of locos languishing at Barry had already gained momentum. Many of the later arrivals there had been of the BR Standard classes, but some of the longer-term residents were of more exceptional types, such as for example the two S&D '7F' 2-8-0s, three 'Crab' 2-6-0s, plus examples from the GWR 'Castle' and 'King' Class 4-6-0s, and of course the singularly unique BR Standard '8P' Pacific 71000 *Duke of Gloucester*. And so, in September 1968, the first of over two hundred locos to eventually escape Barry was '4F' 0-6-0 43924, going to the Keighley & Worth Valley Railway, where it remains today. And over the following two decades a total of 213 locos were rescued, sold on into preservation at the equivalent scrap metal price dictated at the time of its purchase from Woodham being made. The Barry scrapyard episode, lasting for almost three decades, is a fascinating chapter in British steam loco history, having made such a profound influence on the foundation and success of our much loved and visited heritage railways of today.

Opposite page: A scene at Dai Woodham's Barry scrapyard on 19 July 1968, just days before the end of steam traction on BR. In the foreground were two LMS Ivatt '2MT' 2-6-0s, 46428 and 46447, latterly of Crewe South and Wigan Spring's Branch depots, respectively.

Right: A glimpse at Barry scrapyard on 7 November 1970, through the fireman's side spectacle plate of '3F' 0-6-0 tank 47298, withdrawn from Sutton Oak MPD in December 1966. (Both author's collection)

Images of locos being cut up at Barry scrapyard are indeed rare, and in this April 1972 scene little was remaining of this former Wigan Spring's Branch depot's BR Standard '4MT' 2-6-0 76080, having already been reduced to more manageable pieces of scrap metal. Just one further steam loco was scrapped at Barry during the 1970s, GWR '28xx' Class 2-8-0 3817, which was dealt with in 1973. (Kevin Lane)

Above right: Some of the earlier arrivals at Barry were scrapped in relative haste, and this 1964 photo of 'Prairie' 5547 shows it still bearing its cast smokebox numberplate, unlike most later arrivals at Barry. The GWR '4575' Class 2-6-2 tank had been built at Swindon in 1928 and was taken out of service at Swindon's depot in February 1962. Its final duties included working the Cirencester-Kemble branch line passenger services. Notable for this period, it was not disposed of at Swindon Works, possibly due to the accumulating workload at that time in its Concentration Yard ('Con Yard'), consequently being dealt with at Barry eventually during March 1965. With so many locos having been rescued from scrap at Barry, many having been returned to steam after several years of dedicated fundraising and painstaking restoration work by various groups and societies, with just a small number probably unlikely ever to see action again, it is easy for one to overlook the fact that sixty-four, of mainly tank loco classes, were disposed of there between 1959 and 1965, including representatives from classes '1366', '3150', '4500', '4575', '5101', '5400', '5600', '5700', '6100', '6400', '7200' and '9400', plus eight '4300' Class 'Moguls'. Additional notable disposals received from SR depots were two 'S15' 4-6-0s, 30512 and 30844, from Feltham, Bulleid Light Pacifics 34045 *Ottery St. Mary* and 34094 *Mortehoe* from Bournemouth and Nine Elms respectively, Ivatt '2MTs' 41248 and 41303 from Barnstaple Junction and Bournemouth respectively, BR Standard '4MT' 80067 from Templecombe, and '9F' 92232 from Cardiff East Dock, all disposed of during 1965. The BR Standard '9F' 2-10-0 had scandalously seen just six years and four months of BR service!

Below right: A one-time GWR/WR top link 'Castle' Class 4-6-0, 5080 *Defiant*, formerly a South and West Wales-allocated loco at Cardiff Canton, Landore, Carmarthen and finally Llanelly, the latter from where it was cast aside in April 1963, languishing in the salt-laden air at Barry in 1966. Minus its tender, like many other locos eventually found themselves, and sandwiched between 4953 *Pitchford Hall* and '5101' Class 'Prairie' tank 5164, it had been given some degree of dignity, with its former running number chalked in the scar left where its gleaming brass cab-side numberplate had once been prominently affixed. In August 1974, it was purchased by the Birmingham Railway Museum Trust (now Vintage Trains) as a spare components donor for 7029 *Clun Castle*, but living up to its name, it was by 1988 remarkably to be found in a fully restored main line-certified condition again. (Both author's collection)

Above left: BR Class '41' D601 *Ark Royal* at Barry in June 1973. Built by the North British Locomotive Company, it entered BR service in March 1958, one of a class of just five 2,000hp diesel-hydraulics. Initially used on the Paddington-Plymouth-Penzance route, the Class '41' 'Warships' soon became restricted to services west of Plymouth, but before long BR considered such a small class non-standard, so it was inevitable that they were early withdrawal candidates. Allegedly suffering with reliability issues, they were taken out of service en masse at Plymouth's Laira depot in December 1967. Green-liveried D601 and sister loco D600 *Active*, already in the later BR blue livery, arrived at Barry shortly after withdrawal. The three other class members were hastily disposed of at Cashmore's Newport yard during August 1968. D600 was cut up in March 1970, but D601 somehow survived until June 1980, increasingly as a vandalised hulk, along with another Barry resident, North British-built diesel-hydraulic Class '21' D6122.

Below left: As previously related, many locos purchased by Woodham for scrap never escaped. One such unique acquisition was Hunslet 'Austerity' 0-6-0 saddle tank *WD 203*, photographed there in August 1963 shortly after its arrival from the Longmoor Military Railway in Hampshire, where it had seen very limited service. Following a period of 'strategic storage,' it was deemed surplus in July 1963 and despatched to Barry for disposal. The War Department took delivery of 377 of the Hunslet-design 'Austerity' locos, supplied between 1943 and 1947 from six established loco builders. Many became surplus after hostilities ended and were sold on into industry. In 1952–53, the Ministry of Supply received a further fourteen (Works Nos 3790–3808, allocated fleet Nos *WD 190–203*) to the same successful design, and built on the same Hunslet production line as further 'Austerity' locos supplied to industry. These were to be the last steam locos used by the British Military, *WD 203* being the final 'Austerity' saddle tank delivered, and becoming the first to be disposed of, just ten years old. The last regular working example was to be *WD 192* (later 92) *Waggoner*, which had seen use at the Marchwood Military Port until as late as 1982. (Both author's collection)

Above left: A poignant scene of Bulleid Pacifics and BR Standard locos rusting away at Barry Docks in August 1979. In the foreground was 'Merchant Navy' Class 4-6-2 35025 *Brocklebank Line*, revealing its asbestos lagging, having already stood in the open there for some fourteen years. Rescued in 1986, it eventually passed into the ownership of Southern Locomotives Ltd and has been the subject of a long-term restoration project at Sellindge in Kent. (Author's collection)

Above right: A row of GW 'Prairie' tanks and 'Manor' 4-6-0s at Barry in August 1979, with 'Small Prairie' 5538 the closest of the tank locos, still, but only just revealing its original-style BR 'cycling lion'. (Author's collection)

Right: Bulleid Light Pacifics, rebuilt 'West Country' 34046 *Braunton* and un-rebuilt 'Battle of Britain' 34070 *Manston*, embrace at Barry on 22 August 1970. They were subsequently rescued for preservation, and both have returned to active service, 34046 as a main line performer in the care of the Locomotive Services Group at Crewe, and 34070 maintained on the Swanage Railway in Dorset. (Roy Burt)

Left: A 1966 overview of but a small sector of the major of the two yards at a time preceding the rescue of most locos, which had found their way to Barry, and with yet more to arrive. Woodham's yard proved to be a saviour for Britain's heritage railways of today, with coaches and wagons luckily taking precedence over locos for early disposal at Barry.

Above: A scene at Barry in August 1966, with GWR Collett '2887' Class 2-8-0 3862, formerly of Croes Newydd MPD, and Collett '5101' Class 2-6-2 tanks 4156 and 4110 formerly of Severn Tunnel Junction depot. 'Prairie' 4156, along with BR Standard '9F' 92085, was cut up at Barry in summer 1980 when the supply of condemned wagons from BR dried up, however 3862 and 4110 happily survived into preservation.

Opposite page above left: Maunsell 'S15' Class 4-6-0 30825 at Barry in the mid-1960s. Certified to work Whitby trains, it is one of three 'S15' Class locomotives rescued from Barry and owned by the Essex Locomotive Society, all of which are kept at the NYMR. (All author's collection)

Above right: BR Standard '9F' 2-10-0 92214 minus tender, and Collett 'Heavy Goods' 2-8-0 3850 at Barry on 19 July 1968. 92214 was withdrawn from Severn Tunnel Junction MPD in August 1965 after just six years of BR service. Rescued in 1980, it has since been restored to full working order, more recently working on the Great Central Railway (GCR). 3850 was withdrawn from Croes Newydd MPD in August 1965, and has also been restored to working order, enjoying service on both the GCR as well as on the Gloucester & Warwickshire Railway.

Below right: BR Standard '4MT' 2-6-4 tank 80136 at Barry in the mid-1970s, by then earmarked by the Flint & Deeside Railway Preservation Society for use on the Llangollen Railway. Following restoration at the Churnet Valley Railway in 1998, it worked both there and on the Llangollen Railway until the expiry of its boiler certificate. It was latterly moved to the NYMR and returned to service on this popular Moors Railway during 2016. (Both author's collection)

Left: A diverse line-up of former GWR Churchward and Collett-designed classes at Barry in early 1967: 2-8-0 2857, 2-8-0 tank 5239, 2-6-2 tank 5199, 2-8-0 2874, 2-6-2 tank 5164, 'Castle' Class 4-6-0 5080 *Defiant*, 'Hall' Class 4-6-0 4953 *Pitchford Hall*, and 2-8-0 tank 5224. All except 2874 have since returned to steam.

Opposite page: SR veterans hastily being disposed of at Eastleigh on 6 July 1961, with 'E2' Class 0-6-0 tank 32108 being dealt with, and 'D1' Class 4-4-0 31735 next in line. The Billington 'E2' had been withdrawn earlier that month from Southampton Docks depot, and the Maunsell 'D1' from Eastleigh in April 1961. (Both author's collection)

Right: BR Standard '9F' 2-10-0 92085 at Barry on 5 August 1967, having arrived four months before. Built in 1956, the single-chimney '9F', one of just two so fitted to arrive, had been taken out of service at Birkenhead in December 1966, with a working life of just ten years. Despite this, it was unable to escape the cutter's torch, and met its fate at Barry in July 1980 and possessing the dubious honour of being one of the last two locos to be scrapped there, that irreversible deed being undertaken in the company of former Severn Tunnel Junction's '5101' Class 'Prairie' 4156. (Author)

3

DISPOSAL AT BR WORKSHOPS & DEPOTS

Above left: Derby Works in June 1963, with '3F' 0-6-0 tank 47413, formerly of Lostock Hall MPD, and '4F' 0-6-0 44129 formerly of Staveley (Barrow Hill) in their final days before being broken up later that month. An unidentified BR Standard Class '5MT' 4-6-0 was also present, although the first members of this class were not officially taken out of service until 1964.

Above right: A railfan's group visit to Darlington North Road Works on 16 November 1963. Class 'J21' 0-6-0 65033 was subsequently saved, but it was to be a different outcome for the North British diesel-hydraulic shunter D2700 from Goole depot, which was cut up in September 1964 after just nine years of service.

Left: Gresley 'A3' Pacific 60082 *Neil Gow* and BR Standard 'Clan' Class Pacific 72003 *Clan Fraser* awaiting the torch at Darlington Works on 26 October 1963. (All author's collection)

Awaiting their call for breaking up, LMS Fairburn '4MT' 2-6-4 tanks 42100, previously of Watford depot, and 42117, previously of Willesden depot, stand dejected in the yard at Derby Works early in the 'Big Freeze' of 1963. Both had met their fate during the June of that year, by then having long been thawed out. (Author's collection)

Above left: Power car (non-driving) Sc49004 from the ill-fated 25kV electric Class '370' Advance Passenger Train Prototype (APT-P) project awaiting disposal at Derby Litchurch Lane works on 15 February 1986. First officially unveiled in June 1978, the pre-production APT-P passenger train units were used on the West Coast Main Line (WCML) between London Euston and Glasgow Central. Three complete sets were built by British Rail Engineering Ltd (BREL) at Derby Works between 1977 and 1980. A complete APT-P train comprised two units, each of seven cars, and ran in a back-to-back formation with the two motor cars adjoining each other. The motor cars had no seating accommodation or through-gangway, making the two halves of the train divorced, to the inconvenience of passengers and train staff. The eight traction motors of the two central motor cars gave a total output of 8,000hp, enabling the tilting train in December 1979 to attain a then UK top rail speed record of 162.2mph. Testing of the APT-P continued until 1986, but repeated technical issues and financial cut-backs saw the costly project abandoned, attracting negative media attention and controversy. Although the production series APT-S (Squadron Service) trains were never built, the later InterCity 225 ECML passenger vehicles and their related Class '91' locos, as well as the 'Pendolino' Class '390' trains built for use on the WCML, were successfully developed using the experience gained from the project. All six APT-P units were taken out of service during 1985–86, and thirty-seven of the original forty-four cars were rapidly disposed of in relative secrecy, despatched to Booth's scrapyard in Rotherham. Six surviving cars from unit 370003 were retained by BR and are now exhibited at the Crewe Heritage Centre.

Below left: Former BREL Crewe Works 1972-built prototype High Speed Train (HST) power car 41002, latterly used as departmental ADB975813, discarded in the yard at Derby Works on 15 February 1986. The two original prototype Class '41' HST power cars 41001 and 41002 were used on tests on the ECML and GW main line, latterly being reclassified in the carriage numbering range W43000 and W43001, respectively, when the production HSTs were introduced. In November 1976, they were taken out of capital stock and used by the Railway Technical Centre (RTC) at Derby as 'dead load' vehicles, accordingly renumbered ADB975812 and ADB975813, respectively. The two former prototype HST power cars were used in APT-P and InterCity 225 test trains. Both power cars were taken out of departmental service in 1983 and disposal of W43001 took place at Booth's Rotherham scrapyard during December 1990. 41001 was more fortunate, and was saved for display at York's National Railway Museum. (Both Author)

Right: Partly dismantled Class '40' 40158 at Doncaster Works on 9 September 1984. Introduced to BR service at Haymarket Traction Maintenance Depot (TMD) in September 1961, and loyal to the Scottish Region for some two decades, it ended its days between May 1981 and December 1983, based at Carlisle Kingmoor TMD. Seven members of the original total of 199 class members survive in preservation today. (Roy Burt)

Left: On 27 July 1980, Class '31/0' 'Toffee Apple' Brush Type 2s 31015 and 31006 were awaiting the cutter's torch at Doncaster Works, an act which came about during November and December of that year, respectively. These pilot scheme locos had spent their entire working careers based at Stratford depot, rarely straying far from their home territory. 31015, withdrawn from service in May 1980, was originally delivered new from the Brush Falcon Works to Stratford depot in July 1958 numbered D5515. D5506 was also delivered new to Stratford, in March 1958, and was taken out of service as 31006 in January 1980. They were recognisable by not having the route indicator box mounted on the roof above the cabs, and the 'Toffee Apple' nickname by train crew came about by the shape of the single removable power handle used in both cabs. As non-standard locos, D5500–D5519 were fitted with red circle electro-magnetic control equipment, thus sealing their fate somewhat earlier than the sister Class '31/1s'; however, following a collision, D5518 was rebuilt in 1967 as a standard Class 31/1 with blue star coupling code, renumbered 31101 and survives in preservation today. (Author)

Above left: By 12 May 1985, 'Peak' Class '46' 46032, delivered new to Gateshead depot from BR Derby Works in June 1962 as D169, had been 'disembowelled' at Doncaster Works. The cab of Class '03' shunter 03175, previously dealt with in November 1983, was being used as a convenient staff mess hut.

Above: With the ongoing rationalisation of BR's railfreight services throughout the 1970s, and the consequent redundancy of many Class '03' 204hp shunting locos, over thirty such examples were disposed at Doncaster Works during the late 1970s and early 1980s, in addition to those similarly dealt with at Swindon Works and by some private contractors. Some '03s' never survived to receive their official TOPS designation, and Swindon Works 1960-built (D)2173, latterly from Hull Botanic Gardens depot, was one such example, withdrawn in November 1973 and noted at Doncaster Works on 23 May 1976. Receiving the TOPS numbering indeed was no assurance of longevity, for former Bradford Hammerton Street depot's 03091 had also been taken out of service, in March 1974. Both were eventually reduced to scrap metal during the following March. (Both Roy Burt)

Above: The disheartening hulk of a once greatly idolised ECML 'racehorse', 'Deltic' Class '55' 55012 *Crepello* was captured during its final hours at Doncaster Works in September 1981. (Author's collection)

Opposite page below left: Class '13' 'Master & Slave' 13003 (formerly D4500) at Doncaster Works on 12 May 1985 was the progenitor of three such units rebuilt from Class '08' shunting locos at Darlington Works in 1965 to work the Tinsley marshalling yard hump. 13003 was disposed of at Doncaster in September 1986. (Author)

Above: The first 'A1' Class Pacific withdrawal was that of Doncaster Works 1949-built 60123 *H.A. Ivatt*, which had experienced a mishap at Offord on the ECML near Huntingdon. Working its first recorded non-passenger train, a Kings Cross Goods to Ardsley freight, it was in collision with another in the down goods loop on 7 September 1962. It was photographed at Doncaster Works on 13 October shortly before its disposal. (Industrial Railway Society – Kevin Lane)

Opposite page: On 28 March 1979, three years following its withdrawal from service, some cab remnants of former Stoke Division's Sulzer Class '24/0' 24064 were still scattered about in the Swindon Works 'Con Yard' following an appointment with the busy cutter's torch in December 1976. (Kevin Lane)

MCO
B571546
M.C.O.
COND
75014
24 064

Right: The driver's cabs of 'Western' Class '52' diesel-hydraulics D1001, formerly named *Western Pathfinder*, withdrawn from Laira TMD in October 1976, and D1065, formerly named *Western Consort* and withdrawn from Laira in November 1976, were grounded at Swindon Works on 28 March 1979. Both 'Westerns' had been cut up during August 1977. (Kevin Lane)

Left: The scrapmen eagerly get to grips with a driver's cab of Beyer Peacock Class '35' D7058 at Swindon Works on 20 June 1972. This November 1962-built 'Hymek' had completed less than nine years in traffic, working out of Cardiff Canton and Old Oak Common TMDs, and had been taken out of traffic at Cardiff in October 1971, the first year of official withdrawal of members of this class, numbering eighteen during that year. None of this popular mixed-traffic diesel-hydraulic design, numbered D7000–D7100, received their allocated Class '35' TOPS numbers. Regarded by BR as 'non-standard' due to their hydraulic transmission, the last 'Hymeks' were switched off during 1975. (Kevin Lane)

Opposite page: 'Western' Class '52' D1054 *Western Governor*, in the later BR blue livery, and D1001 *Western Pathfinder* behind, were awaiting their destiny at Swindon Works on a wet day in early 1977. D1054 had been withdrawn from Plymouth Laira TMD during the previous November, and was soon reduced to scrap metal during May 1977. (Author's collection)

Above left: Condemned Beyer Peacock diesel-hydraulic 'Hymek' Class '35s' at Victoria Road sidings, Bristol, on 13 January 1972. Left to right in the front of the three rows were D7041, D7063 and D7040, awaiting their movement to Swindon Works for breaking up, which took place later that year. A total of 101 'Hymeks' were introduced between May 1961 and February 1964, and BR's short-sighted decision to eradicate diesel-hydraulic locos cut short their successful careers. With withdrawals taking place between 1971 and 1975, most were disposed of at Swindon Works, but a small number were dealt with by private contractors. Some saw less than ten years of service, but two survived into the 1980s based at the Mickleover RTC facility, denigratingly used as dead load vehicles. Happily, four of this well-liked and reliable class survived into preservation. (Roy Burt)

Below left: A poignant scene outside the Swindon Works erecting shop on 29 April 1982. Waiting to enter for internal component removal were Class '40s' 40117, withdrawn on 7 September 1981 and finally scrapped in November 1983; 40175, withdrawn on 31 May 1981 and scrapped in February 1983; and 40098, withdrawn on 14 April 1981 and reduced to scrap during April 1983. A scandalous decision at the time, the works was to remain open for just four further years. (Roy Burt)

Opposite page above left: A visit to Swindon Works on 5 August 1970 found two pilot scheme 'Warship' Class '42' diesel-hydraulic locos, maroon D801 *Vanguard*, and blue-liveried D802 *Formidable*, both discarded following stripping of any useful components to aid some remaining operational members of the class. Rated at 2,000hp, instead of the standard 2,200hp for the remainder of the class, D800–D802, as non-standard examples, were to be early candidates for withdrawal. Indeed, it is remarkable that D802 had even been treated to the corporate BR blue livery, which it could not have borne for long. These were the first two class members scrapped at Swindon, the deed having been completed during the October of that year. 'Warship' Class progenitor D800 *Sir Brian Robertson* had been dealt with at Cashmore's Newport yard during the previous year. These two pilot scheme late 1958-built locos, D801 and D802, had a working life of less than ten years, withdrawn from Plymouth's Laira TMD in August and October 1968, respectively. Noteworthy is that between March 1969 and December 1972 all remaining 'Warship' Classes '42' and '43' locos, which had originally been based on the successful German DB 'V200' Class design, had been taken out of service. (Author)

Right: Class '42' diesel-hydraulic D818 *Glory* hidden away inside the weighbridge building at Swindon Works on 5 October 1980. Built in 1960 at Swindon, D818 had been taken out of BR service in November 1972, but survived the cutter's torch merely due to its selection as a source of spares for sister loco D832 *Onslaught*, claimed by BR's Research Department at Derby. Its longevity led to the belief that it had an assured future and was the subject of a preservation project, particularly with external attention being lavished upon it by the Work's apprentices. Indeed, it was repainted back into green from the corporate BR 'Rail Blue' livery, and had been openly displayed around the external turntable adjacent to the main works entrance. Soon after the impending works closure announcement, seemingly contemptuously timed to coincide with the 150th anniversary of the GWR, D818 was hastily broken up, leaving insufficient time for the launch of a preservationist's rescue. Very sadly, the works was duly closed on 26 March 1986, but by way of some small compensation, the historic GWR 1906-built weighbridge building subsequently survived to see use as a micro-brewery, bar, and restaurant. (Author)

Above: The falling rain and autumn leaves exemplify this sombre scene of '03' Class 204hp shunter 03142 and Class '40' 40112 awaiting the cutter's torch at the Swindon Works 'Con Yard' on 29 November 1984. 40112 was one of the small batch of twenty English Electric Type 4s, numbered D305–D324, to be built at the Darlington Works of Robert Stephenson & Hawthorns Ltd, and was delivered new to Crewe depot in December 1960, numbered D312. It was reduced to scrap metal at Swindon during October 1985, at the same time as 03142, one of just six class members which subsequently received cut-down cab conversions, enabling them to exclusively work the coal traffic on the loading gauge-restricted Burry Port & Gwendraeth Valley line in Carmarthenshire. (Author)

Above left: Derby Works 1957-built Class '08' 08328 and two former steam loco tenders, a Collett 4,000-gallon type and a former Railway Operating Department (ROD) 4,000-gallon tender numbered 1989, subsequently modified with a reduced height. Both were pending the cutter's torch at Swindon Works on 31 October 1983. The ROD tender had originally been paired with North British 2-8-0 (Works No.22047 built in 1918), which had served on the Western Front in France during the First World War. The GWR acquired around one hundred 2-8-0s after the hostilities ended and those were numbered in the '30xx' series. As the ex-ROD locos were gradually taken out of service by the GWR, some of their tenders continued to see further useful service. After receiving light repairs and the fitting of vacuum brakes, they were paired with 'Aberdare' Class 2-6-0s, and latterly the Collett '2251' Class 0-6-0s. This tender was quite a remarkable survivor by this time, it quite probably having found use in its later years as a water carrier, designated for internal use only at Swindon Works. The 350hp shunter was built at BR Derby Works and was taken into traffic at Cardiff East Dock in November 1957, numbered D3398. It remained on WR books until October 1967, when a transfer took it to the North West at Speke Junction MPD. It was withdrawn from service at nearby Allerton TMD on 31 December 1981, and it was not until January 1987 before it was finally cut up. (Author)

Above right: Class '31/1' 31241 was in the process of being dismantled at the Swindon Works 'Con Yard' on 29 April 1982. The Brush Traction-built Type 2 had been accepted into service at March depot in November 1960, numbered D5668. Withdrawn from Old Oak Common TMD in March 1982, the completion of its disposal at Swindon was rapid, the task having been accomplished by May 1982. (Roy Burt)

Below: On 29 April 1982, the hulk of Class '08' shunter 08230 stood at the head of a row of condemned Class '25/2' and Class '40' locos, with 25137 leading, all awaiting their fate at the 'Con Yard'. The heavily stripped 350hp shunter had been built at BR's Darlington Works, delivered new to Stratford MPD in April 1956 numbered D3300, and where it remained until March 1989 as a member of the depot's large fleet of diverse classes of shunting loco. It was ultimately moved on to March TMD to see out its time, its withdrawal date arriving during September 1980. It was eventually broken up at Swindon in May 1983, closely followed by 25137 during the following month. (Roy Burt)

Above: The decrepit remains of Class '31/1' 31111, 'Peak' 46004, and 350hp shunter 08192 cut a solemn picture as they awaited their fate in the rain at Swindon Works on 29 November 1984. Built at the Brush Falcon Works and introduced into BR traffic at March TMD in May 1959 numbered D5529, the hulk of the Brush Type 2 was eventually disposed of during July 1986. The BR Derby Works-built 'Peak' had entered traffic at Derby depot as D141 in November 1961, and did not have long to wait on 'death row', being disposed of during March 1985. The veteran shunter, new to Llanelly MPD in July 1956 as 13262, and subsequently renumbered in June 1961 to D3262, had to wait for ten further weeks before the call eventually came from Swindon's fully occupied resident 'Grim Reaper'. (Author)

The scrapman had been busy, with 08045 on 29 November 1984 found to be partially dismembered, the job finally completed by 15 December. 08045 was delivered new from Derby Works to Wellingborough MPD in September 1954, numbered 13058, subsequently D3058, and was finally taken out of service at Toton TMD in July 1982. Alongside it were the remains of a sister 350hp shunter with a dual-braked Class '27/0', 27034, awaiting its turn beyond. Numbered D5382, the BRCW Type 2 had entered service in April 1962, one of the final batch of Type 2s, D5379–D5415, all initially delivered to Cricklewood East depot painted in an appealing two-tone green colour scheme, as also applied to the later Class '25s'. After its initial sojourn in North London, D5382 was to spend most of its working life in Scotland, until its withdrawal beckoned in July 1984. (Author)

Above left: Lined up in the rain alongside Swindon Works on 29 November 1984, awaiting their call for disposal, were 'Peak' Class '46s' 46007, 46006, and 46039, in the company of Class '25/2' 25152. BR Derby Works-built 46006 and 46007 had both entered service at Derby depot in December 1961, numbered D143 and D144, respectively, and 46039 had been delivered new to Gateshead depot from Derby Works in August 1962, numbered D176. The official disposal dates at Swindon's 'Con Yard' were 22 June 1985 for 46007, 24 August 1985 for 46006, and 3 August 1985 for 46039. The Class '46s', numbered D138–D193, were built at Derby Works and introduced between November 1961 and January 1963, the final version of the 'Peaks'. Although based on the ten original Class '44s' ordered in 1955, the '46s' were fitted with Brush electrical equipment. Seeing widespread use on the network, their four-axle bogies and restricted route availability would eventually influence their downfall. The first class member, 46005, was withdrawn from Laira TMD in December 1977 and cut up at Derby during the following March. The combined introduction of the InterCity HSTs, contraction of freight business, and new loco replacements saw their storage at Swindon gather momentum, but immediate disposal for some was not envisaged, for an increased demand for Type 4 traction in 1981 found some better examples reinstated, including 46006 and 46007, but in the event, this proved to be a precipitate move, for these two were officially withdrawn just two months later, in January 1982. The demise of the Class '46s' soon transpired, for the seven remaining members were finally taken out of traffic at Gateshead TMD in 1984.

Below left: Strategically placed at Swindon Works for disposal, 46020 was in the process of being torched at the western extremity of the 'Con Yard' in the persistent rain of 29 November 1984. Built at BR Derby Works and introduced into service as D157 in March 1962, it had been moved from Laira and placed into storage at Swindon in October 1980 pending possible reinstatement, but it was withdrawn at the end of that year after just eighteen years of service.

Opposite page: The last day of existence for Class '25/2' 25215, when in the process of being dismembered at Swindon on 31 October 1983, just four months after its withdrawal from service. Built at Derby Works and delivered new in December 1965 to BR's Nottingham Division as D7565, it saw work in the Midlands, London, Scotland, South Wales, and finally in the West Country, before being taken out of service at Laira TMD in June 1983. (All author)

Above left: Swindon Works on 31 October 1983, with 25184, 31111, 40083, 40154, 40176, 08583, and 40149 all lined up for disposal. 25184 was despatched in February 1984, and the others between September 1985 and July 1986. (Author)

Above right: Derby 1958-built Class '125/2' diesel-hydraulic DMU E51159 pending disposal at Swindon on 5 October 1980. The motor brake second (MBS) had been relieved of its two Rolls-Royce 238hp engines to keep remaining St Pancras-Bedford Class '127' 'Bedpan' units in traffic until completion of route electrification. The three GWR Collett tenders alongside, two 4,000 gallon and one 3,000 gallon, were also for disposal. (Author)

Left: Class '25/3' 25263, Class '25/1' 25070 and Class '31/0' 31017 awaiting disposal on 29 April 1982. Cut up at Swindon in January 1983, 31017 was one of just two Class '31/0' 'Toffee Apples' to be dealt with there. The two Class '25s' followed during May 1983. (Roy Burt)

The wretched hulks of English Electric Type 4s, later Class '40s', 40149 and 40176 had been stripped to mere rolling shells by the time of this visit to Swindon Works on 29 November 1984. Both built at the English Electric Vulcan Works, they had entered service in June 1961 at York North, and in February 1962 at Camden depot, numbered D349 and D376, respectively. Condemned in August 1981 after a prolonged period of open storage at Swindon, they ultimately met their fate in May 1986 and December 1985, respectively. (Author)

Above left: On 29 April 1982, Class '40s' 40016 *Campania* and 40014 *Antonia* await an inevitable outcome at Swindon Works, which arrived in December 1983. These, and other early build and eventually named class members were WCML stalwarts upon their first introduction into BR service, among other duties, ousting the much-loved 'Princess Coronation' Pacifics and other steam classes from the principal express passenger services working out of London Euston to Birmingham, the North West, and Glasgow. 40016 had been delivered new to Crewe North MPD from the Vulcan Foundry in June 1959, numbered D216, and it was to work its entire life in the North West, variously allocated to Carlisle, Allerton, Stoke and BR's Liverpool Division, the latter from where it was taken out of service in May 1981. D214, later 40014 under the TOPS scheme, had been delivered new to Willesden MPD in June 1959. It had also spent its entire career on the London Midland Region (LMR), variously working from Longsight, Edge Hill, Crewe, Camden, Bescot, and finally Longsight depot, from where it was withdrawn from service in November 1981. The Type 4 design had been based on experience gained and lessons learned from the experimental LMS-design Derby Works/English Electric locos 10000 and 10001, and the SR prototypes 10201–10203, all successfully put through their paces up until 1957. Based upon this working experience, BR placed an order with English Electric for ten Type 4 prototypes, D200–D209, which were put to work on Great Eastern and Great Northern Region lines out of London. Following successful 'front-line' passenger service, a series production order was placed, and were built in batches by the Vulcan Foundry at Newton-le-Willows, and at Robert Stephenson & Hawthorns' Darlington Works, both part of the English Electric Company. Delivered between May 1959 and September 1962, numbered D210–D399, they were allocated to various areas of England and Scotland. At home on express and secondary passenger services, as well as on freight duties, the final nineteen class members in routine service were taken out of BR traffic during 1984, although BR's 'Pilot Scheme' and National Collection designate D200, latterly numbered 40122, was retained for working special duties until 1987.

Below left: Class '40' 40037 and the remains of 08194 seen at Swindon Works on 31 October 1983. Delivered new to Gateshead depot in October 1959, D237 was a former ECML stalwart until May 1977, one of a batch of thirteen Gateshead-allocated English Electric Type 4s that took over some of the ECML Pacific steam duties north and south from Newcastle from December 1959. It was withdrawn in August 1981, having spent its final five years allocated to Carlisle Kingmoor TMD. (Both author)

A poignant scene in the 'Con Yard' on 31 October 1983 as a contractor appears to be in deep thought, pausing for an undoubtedly welcome breath of fresh air and rest from fulfilling his arduous task of reducing what appeared to be a former Class '40' to manageable pieces of scrap metal. The 1959-built 'whistling giant', Class '40' 40037, was standing next in line, awaiting similar treatment, which was fulfilled during the following month. (Author)

4

EXTENDED SERVICE & STATIC DUTIES

Opposite page: Two former Sharp, Stewart & Co. 1886-built Lancashire & Yorkshire Railway (LYR) Barton Wright 0-4-4 tanks, minus their centre driving wheels, had been employed as carriage heating boilers for over five decades, latterly at Cheetham Hill sidings, Manchester. The boilers bore numbers 925 and 903, which suggested their probable former LYR identities as being 519 and 636, taken out of capital stock in 1911 and 1908, respectively. Their premature withdrawal had been influenced by the introduction of the LYR Aspinall 2-4-2 radial tanks from 1889. Having seen such limited use, several 0-4-4 tanks were retained for carriage heating duties at sidings around the LYR system, including at Accrington, Blackpool, Colne, Edge Hill, Irlam, Manchester and Southport, where they remained, with many passing into BR ownership, their boilers periodically maintained at Horwich Works. With their heating duties redundant by the early to mid-1960s, these two veterans were recovered from Cheetham Hill to Newton Heath depot for breaking up, where they were photographed in July 1967 alongside the breakdown crane siding just before their demise. (Charlie Cross)

Above right: On 19 October 1975, A.C. Cars 1958-built railbus Sc79979 was being used as a mess van at Craigentinny carriage sidings. Having worked on various branch lines in Scotland during the late 1950s and throughout most of the 1960s, the railbus spent more time in use as a staff mess and store van than it had in passenger service. In the early 1980s the vehicle body was recovered and transported to the heritage Strathspey Railway, intended for use as staff facilities. However, the removal cost of the blue asbestos contamination eventually influenced the decision to consign it for scrap, and this duly took place during 1992 at M.C. Metals in Glasgow. (Roy Burt)

Below right: Former Southampton Docks shunter, 275hp Class '07' 0-6-0 diesel-electric 2991, one of fourteen built for BR in 1962 by Ruston & Hornsby and numbered D2985–D2998, was the sole survivor from the four withdrawn from service at Eastleigh depot in May 1973, it never having carried its allocated TOPS number. It was subsequently retained as a static load generator at BREL Eastleigh Works and was photographed there on 3 November 1984, by which time it had fallen into disuse. Avoiding being broken up, it was eventually purchased by Knight's Environmental Services Ltd, restored to working order and employed as the works pilot loco, at the by then privatised Eastleigh Works facility. Repainted and displaying the TOPS numbering 07007, which it never carried when in BR service, remarkably this sixty-year-old veteran was still enjoying regular use in 2024, some fifty-one years after BR had taken it out of service. (Author)

Left: Former British Thomson-Houston-built BR Type 1/Class '15' D8233 in its later incarnation designated as departmental ADB968001. Introduced into BR service in August 1960 at Stratford depot, and moving to Finsbury Park in April 1963, it was withdrawn from BR capital stock in February 1969 and converted for use as an ETS generating unit, employed at Leeds Neville Hill, Edinburgh Haymarket, Craigentinny, Finsbury Park, and Colchester. Relieved of its ETS duties, it was photographed out of use at Stratford TMD on 14 August 1982, six months before it was to be moved to Healey Mills yard for storage. Its departmental role fortunately assured its survival, and it was subsequently acquired for preservation by the 'Class 15 Society'. In 2024 it was under restoration at Baron Street depot on the East Lancashire Railway (ELR). (Roy Burt)

Right: Departmental ADB968014 out of use at Yarmouth Vauxhall carriage sidings on 23 August 1982. Introduced into service in December 1957 as Brush Type 2 D5502, it was withdrawn from capital stock at Stratford in January 1980 as Class '31/0' 31002 and converted to an ETS coaching stock pre-heating mobile generating unit, renumbered ADB968014. Taken out of departmental service in October 1982 when there was no further requirement for such duties, it was disposed of at Crewe Works in May 1984. Four redundant Class '31/0' 'Toffee Apple' locos were so converted at Stratford in early 1980, renumbered ADB968013 (ex-D5513/31013), ADB968014 (ex-D5502/31002), ADB968015 (ex-D5514/31014) and ADB968016 (ex-D5508/31008). (Author)

The unique Class '80' 25kV AC electric loco BR E2001, at Akeman Street, Westcott, on 25 January 1970. It had been built by Metropolitan-Vickers as prototype 3,000hp gas turbine loco 18100. Originally ordered by the GWR, its eventual delivery in 1951 had been hampered by the Second World War. BR converted it from gas turbine to electric propulsion in 1958, initially renumbering it E1000, and subsequently E2001. It had been turned out in main line black livery with silver-painted bogies and lining when new, and this scheme was retained throughout its career, even following its conversion to electric propulsion. It was seconded to the WCML 25kV AC electrification project in 1958, and used for the testing of newly installed overhead line equipment and for staff training purposes. Following commissioning of most of the production 'AL1' to 'AL5' Class locos, E2001 was placed into store in 1961, and it languished for over ten years in various locations including at BR's Rugby Testing Station. When photographed adjacent to the secretive Ministry of Aviation Royal Aircraft Establishment (Rocket Propulsion Department), the ensemble was being used to monitor loco pantograph wind velocity, where it was attached and wired up to two former LMS 50-foot 'BG' coaches, which had previously seen use on the WCML electrification project. Upon completion of these tests, E2001 was sold to J. Cashmore in November 1972, duly being dismantled in their Great Bridge scrapyard during January 1973. (Roy Burt)

From September 1964, North British Railway Cowlairs Works 1891-built veteran 'J36' 0-6-0 65234, still officially in capital stock, had been employed as a stationary boiler alongside the ECML at Edinburgh St Margaret's, supplying steam to buildings on the opposite side of the main running lines from the depot, where it was photographed in August 1966. In April 1967, it was finally officially withdrawn from service and was moved across to the MPD, prior to its despatch for scrap, carried out by contractor Motherwell Machinery & Scrap of Wishaw, whereupon it was broken up in August 1967. (Author's collection)

5

SOUTHERN ELECTRIC STALWARTS

Condemned Southern Railway '2-HAL' (Class '402') 2662 and '2-BIL' (Class '401') 2144 awaiting despatch from Gatwick down yard on 4 January 1970. These units had been introduced to operate long-distance semi-fast services on the newly electrified lines from London to principal Sussex and Hampshire coastal towns, as well as Reading. Eastleigh Works 1939-built 2662 had been taken out of service in September 1969 and was moved from Gatwick down sidings on 24 January 1970 to the Bird Group at Long Marston for scrapping. Eastleigh Works 1938-built 2144 had been withdrawn from service in October 1969 and was moved from Gatwick on 17 February 1970 for scrapping by Armytage & Sons Ltd at Sheepbridge. (Author)

Above left: Sidings near the rural Hampshire village location at Micheldever were used as a holding point for SR rolling stock, pending reformation or disposal, and was established alongside a former wartime fuel and oil terminal constructed in 1939. On 10 January 1971, awaiting disposal was unique '6-TC' unit 601 in the company of '4-SUB' unit 4292. Eastleigh Works reformed the '6-TC' in January 1965, for trialling push-pull equipment for the Bournemouth electrification scheme, using two '4-RES' motor coaches and four intermediate trailers from disbanded '6-PUL' and '6-PAN' units. Class '33' D6580 was also modified for remote operation from 601's driving cab. Used from January 1966 on the London Bridge-Oxted line peak-time services, upon completion of the trials, its front cab corridor connections were removed, the ends plated over, and it was repainted from green to blue livery and transferred in July 1967 to the Clapham Junction-Kensington Olympia 'Kenny Belle' diagrams. Its survival was short-lived, for it was taken out of service after incurring accident damage at Kensington Olympia in July 1970. Condemned in October 1971, it was subsequently dismantled at Brockenhurst sidings during March 1972. As a result of the trial, push-pull '3-TC' and '4-TC' Mk1 units were subsequently introduced on the newly electrified Bournemouth line, and a further eighteen 'Cromptons' were converted to push-pull operation, designated Class '33/1'. (Author)

Below left: The victim of an unfortunate accident at Ramsgate depot in January 1966, the shell of MBS S14544S of '2-HAP' unit 5624 was at Micheldever on 31 August 1969, before cutting up on site in October. The BR standard design '2-HAPs' were first introduced on the South Eastern Division in 1957–58, later to be found on all three SR Divisions. Designed for main line services, they were equipped with express gear ratios. Using reclaimed underframes of 1930s '2-NOL' units, the first batch of thirty-six units, numbered 5601–5636, were built to the older Southern Railway Bulleid-style design. (Roy Burt)

Opposite page above left: SR departmental unit S15 was formed from a '4-SUB' unit utilising driving motors S8420S (departmental DS70057) and S8473S (departmental DS70058), the '4-SUB' 'Mary' units having previously utilised '3-SUB' vehicles of 1925 Southern Railway origin. Following war damage, they were reformed during 1945–46 using spare trailer cars, emerging as '4-SUBs' 4300-4354, some Bulleid-designed trailers used in the sets giving them an odd appearance. The Southern Railway lost a total of ninety-three EMU vehicles during wartime enemy action. All the EMUs in the series 4300–4354 were withdrawn between 1959 and 1962, after which departmental unit S15 was reformed. This 22 March 1970 photo shows it awaiting disposal from Micheldever, soon after its withdrawal from departmental service. (Roy Burt)

Above right: Departmental unit DS70258-DS70259 at Micheldever on 10 January 1971, reformed from two MBS coaches, S11057S-S11058S, respectively, from 1935-built '6-PAN' unit 3031. Two composite first- and second-class Pullman kitchen cars, *Bertha* and *Ruth*, were placed in the centre of this formation, which had been used on push-pull and conductor line clearance trials on the Bournemouth line. Taken out of service in late 1969, the former '6-PAN' MBS coaches were despatched to A. King's Wymondham yard for dismantling in early 1972, although the Pullman cars survived into preservation. (Author)

Below right: MBS S11003S, originally from '6-PUL' unit 3002, but used in a reformed '6-COR' unit 3046, stood condemned with a brake van attached at Micheldever on 31 August 1969, shortly before its consignment for scrap at T.J. Thomson of Stockton in November 1969. As a short-term-measure the '6-COR' units were reformations from the best of the '6-PUL/6-PAN' units withdrawn from Brighton line services, and saw limited use between 1965 and 1968 on Central and South Eastern Division services due to rolling stock shortages. (Roy Burt)

Above: On 9 July 1983, SR '2-HAP' and '4-SUB' units were at Eastleigh pending internal stripping. On the left, 1963-built '2-HAP' unit 6155 would be despatched after stripping to Swindon Works for bogie recovery, followed by disposal at Mayer Newman, Snailwell, during August 1986. To the right was 1950-built '4-SUB' unit 4660, withdrawn in April 1983. After stripping it was hauled to Vic Berry's Leicester yard in August 1985 for stacking and eventual scrapping. To its right was 1951-built '4-SUB' unit 4733, withdrawn in 1982. After stripping, it was also despatched to Vic Berry's yard, in October 1985.

Left: Partly wrecked in a collision with a diesel shunter at New Cross Gate in May 1970, Southern Railway design '4-SUB' unit 4292, which had been completed at Eastleigh works after Nationalisation in March 1949, was stored at Micheldever following its withdrawal. On 10 January 1971, with some doors already removed from MBSO S10879S earmarked for reuse, the unit was pending despatch to A. King & Son at Wymondham for scrapping, a move which eventually took place in September 1971. (Both author)

Above right: Metropolitan-Cammell in 1932 supplied three umber and cream-liveried five-car all-Pullman units, classified '5-BEL', for *Brighton Belle* services which, apart from a break from public services during the Second World War, plied the route daily between London Victoria and Brighton until their withdrawal from service came on 30 April 1972, by then classified by BR merely as the Class '403' units, and somewhat uninspiringly wearing the corporate blue and grey livery. All units were taken to Brighton's Lover's Walk depot following withdrawal, pending sale by auction and subsequent piecemeal disposal. All but one vehicle from the original fifteen built survive today in one form or another, including a complete set being restored during 2024 to main line condition by the '5-BEL Trust'. One vehicle was lost, *Car No.90* of unit 3052, which was initially moved to Wolverton Works in November 1972 following an expression of interest by Railway Transport Hotels Ltd, and where it was photographed in the late 1970s. It was eventually acquired by the Nene Valley Railway circa 1980, where it received some refurbishment, earmarked for use on its heritage passenger trains. A subsequent move to Carnforth for further refurbishment work saw it tragically damaged by fire there in July 1991. The stricken vehicle was then moved to the ELR, where it suffered further from repeated vandalism and arson attacks at Bury. By 1995 it was declared as being beyond economic repair, and the intractable decision was then made to cut it up on site. (Author's collection)

Below right: Due to a shortage of suburban units following a timetable change in 1969, two 'hybrid' units were reformed at Eastleigh Works, each comprising two 1939-type '2-HAL' MBS units and two '4-SUB' trailers. The units, numbered 4131 and 4132, emerged from the works in May and August 1969, respectively, after Class One overhauls and a repaint in all-over BR blue livery with full yellow ends. Their service requirements proved to be short-lived however, and both units were soon to be taken out of service, on 9 October 1971. On 10 September 1972, unit 4132 was awaiting disposal by contractor A. King and Sons at Wymondham, on the overgrown and truncated stub of the branch that had formerly served Forncett, but was closed to passengers in 1939 and for goods traffic in 1951. This Norwich-based scrap merchant had used this isolated location for the disposal of many condemned BR EMUs, DMUs and passenger vehicles during the 1960s, before stricter legislation for safely dealing with such vehicles containing blue asbestos was brought into force, after which disposal was centred on a purpose-built facility established on a site at Snailwell, near Newmarket (*see pages 108 & 109*). (Author)

6

REGIONAL REJECTION

Right: North British 1959-built Type 2 diesel-hydraulics 6130 and D6137 at Eastfield TMD on 10 August 1969, standing alongside BRCW Type 2 5357. Twenty Class '21s' had their original unreliable licence-built MAN engines replaced between 1965 and 1967 with Paxman Ventura V12 power units, reclassified as Class '29'. Their service was short-lived, declared by BR as non-standard, and with a modest power output of 1,350hp. They were withdrawn in 1971, and both were scrapped during 1972 at Glasgow Works. (Author)

Left: Class '40' 40093 was belching out fumes in the drizzle at Motherwell depot on 23 October 1983, during what was to be its final weeks of BR service. The English Electric Type 4 was delivered new as D293 to Crewe North depot from the Vulcan Foundry in September 1960, and throughout its working life of twenty-three years was allocated entirely to North West area depots. Having sustained power unit damage, it was taken out of service from Wigan (Springs Branch) depot in December 1983 and was disposed of at Doncaster Works during the following May. Class '37/0' 37125 was standing behind, just within the confines of the former stone-built eight-road steam MPD, by then re-roofed and illuminated, but still exuding an element of steam age atmosphere, some sixteen years on from the demise of steam traction at Motherwell, which had taken place in July 1967. (Author)

'Clayton' 900hp Class '17' D8539 at Polmadie depot on 10 August 1969. Delivered new to BR in June 1963 and withdrawn in October 1971, D8539 was cut up at King's Norwich yard some five years later. The Glasgow-based 'Claytons' chiefly worked in the Ayr and Stranraer area, and south via Dumfries to Carlisle. Edinburgh-based locos worked traffic north, but also via the Waverley route to Carlisle. During the late 1960s, several early build examples were moved to Carnforth for banking duties, ousting steam traction (*also see page 11*). The Beyer-Peacock-built batch, D8588–D8616, was initially based either at Barrow Hill or in the North East around Newcastle before reallocation to Scotland. Withdrawals commenced as early as 1968, some having seen barely four years of service. They littered storage sites around Central Scotland following mass withdrawals in October 1971. The Clayton company's ambitious proposal to convert some to battery-electric propulsion for engineering work in tunnel sections never materialised, and they were duly consigned for scrap in Glasgow, Hull, Norwich, and Long Marston. Derby's RTC had acquired problematic D8512, soon replaced by D8598 in spring 1972, and D8521 as a static generator. All were eventually despatched to St Rollox Works, D8512 scrapped in October 1972, D8521 in April 1979, and D8598 in late 1978. D8568 had been sold to Hemel Hempstead-based construction material company Hemelite in 1972, and in 1977 was purchased for use at Clitheroe cement works. This sole remaining class member was acquired for preservation in 1983 by the 'Diesel Traction Group'. (Author)

Above left: Accident damaged Class '20' 20207, and Class '27/0' 27027 awaiting their fate at St Rollox Works, Springburn, on 14 April 1984. English Electric Vulcan Works April 1967-built 20207 was finally cut up in August 1986, and BRCW 27027, built as D5373 in 1962, was dealt with at Thornton yard by contractor Vic Berry in May 1987. (Roy Burt)

Below left: Redundant Wirral & Mersey Class '503' EMUs were being scrapped at Cavendish Road, Birkenhead, on 20 October 1984, with 1938-built DMBS M28685M in the foreground. The 650v DC third rail Wirral & Mersey three-car open sets were LMS-designed and were built jointly by Metropolitan-Cammell and BRCW, both possessing experience in building London Underground stock. Nineteen sets were introduced in 1938, used between Liverpool and the Wirral lines of Rock Ferry, New Brighton, and West Kirby. A second batch of twenty-four sets were built to a similar design by BR in 1956–57. The Class '503s' were an innovation in their day, with air-operated sliding doors controlled by the guard, the first such passenger vehicles so fitted to operate on the British main line. The Class '503s' were progressively taken out of service from June 1983, the final units in passenger service running on 29 March 1985. Scrapping, under contract, took place in various locations, including at Mollington Street depot, Alexandra Dock, BREL Horwich Works, Northwich, and, as here illustrated, at Cavendish sidings on the Birkenhead Docks branch. (Roy Burt)

Opposite page: A line-up of mixed-liveried Manchester-Altrincham Class '505' suburban units outside Altrincham carriage shed on 10 April 1971, during their final service days prior to the route being converted to the standard overhead 25kV AC. The leading motor coaches awaiting the peak services to and from Manchester Oxford Road were M28580M, M28574M, and M28592M. The LMS-designed units, introduced in 1931, were built by Metropolitan-Cammell for the Manchester, South Junction and Altrincham Railway (MSJAR), a route on which they worked exclusively for forty years. Following the 1923 Grouping, the MSJAR company was owned jointly by the LMS and LNER. It operated a 8½-mile route between Manchester London Road (now Manchester Piccadilly) and Altrincham in Cheshire, which was electrified in the early 1930s using the 1,500v DC overhead system. Although assigned by BR to TOPS as Class '505', the units were withdrawn before the numbering system came into common use for multiple units. The 'Altrincham Electrics' were built with a wooden frame construction, most configured with individual compartments without corridors or gangways, and formed as three-car sets, but often operating in multiple in six-car formations during peak travel times. (Author)

Above left: Two of the first batch of five BR 'EM1', later Class '76', locos to be withdrawn from service, 26035 and 26042, awaiting their fate at Reddish depot on 10 April 1971. Both were built at BR's Gorton Works in early 1952 and had been taken out of service together in March 1970. They were cut up on site in late October 1971 by contractor J. Cashmore. Alongside was E76056, formerly named *Triton*, to be taken out of service seven years later. Confusingly, another class member, E26018/76018, illogically received the TOPS numbering 76035 eventually. (Author)

Below left: A line of stored Class '76s', 76002, 76048, 76041, 76046 and 76053, at Reddish depot on 8 January 1983. All were disposed of by Booth of Rotherham during the following twelve months, 76002 being the last to be dealt with in February 1984. (Roy Burt)

Opposite page above right: Following engine repairs, Class '28' Co-Bo D5712 was stabled outside Reddish depot on 26 May 1968. The 'Metro-Vick' was returned to traffic shortly after this, but it was soon to be withdrawn, in the September of that year. It was subsequently cut up at J. Cashmore's Great Bridge scrapyard during December 1969. (Author)

Above right: A little under ten years following its construction by the Yorkshire Engine Company, Class '02' D2869 found itself dumped alongside Newton Heath depot on 10 April 1971. Delivered new to Bank Hall depot in December 1961, it had worked on the extensive dock lines in Liverpool. It was scrapped by Ward's of Beighton during August 1971. These 170hp 0-4-0 diesel-hydraulic shunters were widely supplied to industrial concerns, especially those companies within the chemical and steel industries. BR took delivery of twenty examples during 1960–61, numbered D2850–D2869, and these were built alongside some of their industrial counterparts. Having a short wheelbase, they were ideal for confined dock lines such as at Liverpool, Garston, Preston, and Fleetwood, and most were progressively allocated to Bank Hall, Speke Junction, and Allerton depots during their short BR career. As freight facilities diminished, withdrawals took place between 1969 and 1975, with just three final survivors carrying TOPS numbering. Due to their versatility and having seen limited use, eleven of the class were resold into industry, acquired at knock-down prices compared to newly constructed examples, and seven of the class have subsequently survived into preservation. (Author)

Below right: Class '40' 40046, withdrawn from Healey Mills depot in February 1983, and some BR Mark 1 coaches were used for training exercises at the Ministry of Defence (MoD) base at Moreton on Lugg, near Hereford, in the early 1980s. The depot had its own internal railway system operated by the Royal Army Ordnance Corps and enjoyed a connection with the Welsh Marches line between Hereford and Leominster. During this 1980s period, military special forces and civilian security services were using this establishment to practice tactical entry scenarios onto trains, which necessitated a loco and coaching stock being placed exclusively at their disposal for these discreet and sensitive operations. 40046 performed the 'lead role' between March 1983 and July 1986, when Class '47/4' 47403 replaced it. The damaged coaching stock was rotated on a regular basis, before being consigned for scrap. During this visit on a misty 14 October 1985, 40046 and a Mk1 coach, with its windows quite visibly showing firearms damage, were present along with a former LMS/BR inspection saloon. 40046 was removed from this secretive military base during the following year, and was eventually scrapped at Vic Berry's Leicester yard in June 1987. (Author)

Left: This was the enticing scene immediately visible to visitors at the Crewe Works Open Day on the warm and sunny 22 September 1979. After a modest sum of thirty pence had been paid for an adult admission ticket, these locos could be seen and photographed immediately adjacent to the Flag Lane works entrance used for this event, bordering the Crewe-Chester line. At the head of the row of condemned locos was long-term stored Class '40' 40048. Built at the English Electric Vulcan Foundry and delivered new to Gateshead depot in November 1959 as D248, it had been withdrawn from Haymarket depot in October 1977. It was finally cut up in June 1980, but at Doncaster and not Crewe Works. At the head of the adjacent line of locos to the right and for overhaul was another former Haymarket-allocated Class '40' 40063 which proved to be more fortunate, seeing further service based at Wigan (Springs Branch) depot until its final withdrawal came about in April 1984, one of the last of its class to remain in traffic. Following storage at Bescot depot, it was in June 1987 eventually cut up in Vic Berry's Leicester yard. (Author)

Opposite page below right: At the head of the Flag Lane sidings on 6 March 1988 were Class '47/4' 'Generators' 47409 and 47414. 47409 was delivered new to Finsbury Park in January 1963 numbered D1508, and subsequently spent time allocated to Immingham and Gateshead depots, latterly carrying the name *David Lloyd George*. It was withdrawn from service at Gateshead on 14 August 1986 after sustaining an engine compartment fire at Montrose. Despatched to Vic Berry's yard, it was dealt with during May 1989. It's classmate, 47414, was also delivered new to Finsbury Park TMD from the Brush Falcon Works at Loughborough, in March 1963. It was taken out of service at Gateshead depot on 7 March 1986 after suffering severe underframe damage from a collision with a concrete post which had been placed across the track near York on 3 March 1986. It too was disposed of at Vic Berry's Leicester yard, during June 1989. The buffer stop on the siding upon which the locos were stored would appear to have experienced a heavy shunt incident. (Roy Burt)

Above right: The component recovery area at Crewe Works on 22 October 1983, with Class '84' 84008 having been withdrawn from service some four years previously, on 26 October 1979. After component recovery it was eventually cut up at the works during November 1988. Alongside, 40049 had been taken out of service on 23 January 1983, and was finally cut up at Crewe during October 1985. The more recent arrival of the group, 40183, had been withdrawn from service on 1 June 1983 and was disposed at Crewe three years after its arrival, in June 1986. Sandwiched between the two Class '40s' was another hapless and unidentified class member, by that time already reduced to merely a rolling chassis. (Author)

Below right: Perpetuating the theme of condemned 'whistlers', the scrap line at Derby Works on a snow-covered 15 February 1984 contained the hulks of Class '40s' 40132, 40163 and 40075, and additionally Class '45/0' 'Peak' 45045 *Coldstream Guardsman*. The Class '40s' would eventually be despatched for disposal during January 1987, as part of a short trip working to Vic Berry's Leicester yard. The 'Peak' had already met its fate at the Leicester scrapyard during the previous October. Delivered new from the Vulcan Foundry to Crewe North depot as D332 in February 1961, the split-box 40132 had been withdrawn from service from BR's Liverpool Division during March 1982. Appearing at Derby during the Works open day in September 1982, it had subsequently, along with its companions, become something of a fixture there, languishing alongside the works' perimeter fence for over four years. (Author)

Above left: BR 0-4-0 diesel-mechanical shunter *Departmental No.81* at Doncaster depot in July 1967. Built by Andrew Barclay in 1958 (Works No.424), it entered BR's departmental fleet at its Cambridge Engineer's Depot. Subsequently, transferred to Doncaster depot and taken into capital stock as D2956, it replaced an identical 1956-built sister Barclay which confusingly had carried the same running number, but had been withdrawn in May 1966 at Doncaster depot. Duties by the 'replacement' D2956 proved to be short-lived, for it was taken out of service as early as November 1967, not receiving its Class '01' TOPS numbering. Sold on to Duport Steelworks at Briton Ferry, it was soon broken up in August 1969 and fed into one of the hungry steel furnaces in South Wales. The archetypal D2956 (Andrew Barclay Works No.398), delivered new to BR as 11506, sold on into private industry from Doncaster in 1966, subsequently worked at both the Norwich and Snailwell scrapyards of A. King & Son, the latter location being used for the disposal of rolling stock containing blue asbestos material (*see pages 108 & 109*). There it worked until it was taken out of service in 1984, the yard by then under the ownership of Mayer-Newman Ltd. Its longevity in industry thereby assured its survival into preservation, and this is the D2956 surviving in preservation today, based at Bury in the care of the 'ELR Diesel Group'. (Charlie Cross)

Below left: Life-expired BRCW Class '104/2' DMU set M53487–M53532 at Newton Heath depot on 28 May 1989 had been set aside with serious fire damage, but not visible in this view, on the opposite side of the units. Supplied new to BR in 1957–58, they were cut up on site during October 1990. (Author)

Opposite page: Former LNER 350hp shunter 8001, later BR 15001, stored inside the dilapidated North Staffordshire Railway-built (NSR) roundhouse at Stoke in May 1966. Built at Doncaster Works in August 1944 and briefly used as the works shunter, 15001 moved to March and Immingham depots, and finally to Crewe in January 1966, but saw limited use there. After a brief period in store at Stoke it was officially withdrawn from service in April 1967 and broken up at Cashmore's Great Bridge during August 1967. The NSR built its works in the grounds of an eighteenth-century house circa 1852, and the basement incorporated the tinsmith's shops and iron store. The building, its construction on a scale rivalling that of Camden roundhouse, was latterly used to store condemned locos, but remained derelict into the 1970s when demolished, its historical significance unrecognised outside the local area. (Charlie Cross)

Above: During an era when numerous yards required resident pilot locos, '04' Class D2284, '03' Class 2051 and '08' Class D3692 were stabled at Colchester depot on 2 May 1971, during the 'green to rail blue' livery transitional and pre-TOPs numbering period. Remarkably, all three locos then present at this East Anglian depot have survived; D2284 was to be officially withdrawn from BR service one week later and sold on to the NCB in South Yorkshire, and is now preserved by the 'Heritage Shunters Trust' at Rowsley. D2051 is preserved at the North Norfolk Railway, and D3692 became 08530 and was still extant in 2024, working at Freightliner's Southampton Maritime terminal after extensive refurbishment at Hunslet's Barton-under-Needwood works.

Right: Ipswich depot on 2 May 1971, with non-standard 1960-built British Thomson-Houston BR Type 1s D8230 and 8229, condemned and awaiting disposal. Allocated Class '15' under the TOPS scheme, these numbers were never carried. Both were cut up at Crewe Works, during November 1971 and January 1972, respectively. (Both author)

Opposite page above right: On 26 June 1991, Graffiti-covered 'Peak' Class '45/0s' 45013 (unofficially named *Wyvern*) and 45058 were awaiting disposal, dumped on the site of the former March steam depot, by this time little more than a small workshop building attached to a retaining wall. Built at Derby Works and introduced into BR service at Derby as D20 in February 1961, the 'Peak' was taken out of service as 45013 in April 1987 and disposed some seven years later, during February 1994, by M.C. Metals of Glasgow. 45058 was built at Crewe Works and had entered service at Derby MPD as D97 in April 1961. Withdrawn from service in September 1987, it also was disposed of by M.C. Metals, during March 1994.

Above right: Norwich was BR's last bastion of the 204hp Class '03' 0-6-0 diesel-mechanical shunter, and a line-up of the remaining seven class members were at the Crown Point open day on Saturday 30 May 1987, their final year of BR service. With 03089 just out of view at the end of the row, the remaining six were 03158, 03189, 03397, 03197, 03399, 03084 and 03112, all to be officially withdrawn from service on 5 July 1987, except 03189 which had been withdrawn earlier on 16 March 1986. The final Class '03' member, out of view, would be taken out of service on 6 November 1987.

Below right: Metropolitan-Cammell 'Lightweight' DMUs E79056 and E79272 were stored at the closed Ipswich depot on 2 May 1971, pending disposal. Thirty-six of these two-car 'Lightweight' sets were built, the first railcars procured by BR from an outside contractor. Seven sets were used by the LMR, chiefly on Bury-Bacup services, and the remainder were deployed on rural East Anglian services, although some migrated elsewhere for short periods to Scotland and Lincoln, and for use on the Romford-Upminster shuttle. Metropolitan-Cammell developed what was to become their successful and long-lived Class '101' DMU from this 1955 design. The 'Lightweight' units were distinguishable from the later units by the cowling fitted below the buffer beams and their MU jumper cables. Although a successful product, withdrawal of many rural passenger services in East Anglia under the so-called 'Modernisation Plan' of 1963, as well as their non-standard coupling arrangement, saw their early demise well before life-expiry. All units, apart from one twin set, were taken out of service between 1967 and 1969 and were subsequently scrapped. Two MBS coaches (E79047–E79053) were overhauled at Derby Litchurch Lane Works and converted for use as the RTC's 'Plasma Torch Research Unit', designated as 'Laboratory 21', which saw continued service until 1981, before it too followed the same way as the others. (All author)

Above left: Drewry Class '04' D2276 at Stratford TMD on 7 June 1970, having been taken out of service during August 1969 and never receiving its TOPS numbering. Built at the Darlington Works of Robert Stephenson & Hawthorns Ltd in August 1959, it was to see just ten years of active service with BR, working from Ashford, Norwood Junction and Brighton depots, before a transfer to Stratford in May 1966. Dealer A.R. Adams & Son acquired it, and it was moved to their Newport premises, but was merely used as a spares donor to keep D2244, the company's sister contract-hire loco in operation. Partially dismantled by August 1971, the hulk of D2276 was eventually cut up at Newport in May 1977. The Class '10' D3647 to its right, built at BR Darlington Works in 1958, was deemed as non-standard, being equipped with a Lister-Blackstone power unit, and all class members were early casualties. D3647 was to be cut up by C.F. Booth of Rotherham in September 1970.

Below left: Two BR Sulzer Type 4s, originally D158 and D141, later Class '46' 46021 and 46004, respectively, were reposing at Stratford TMD on 19 November 1983. 46021 was withdrawn from Gateshead depot on 16 January 1983, and 46004 was taken out of service at Plymouth Laira depot on 26 June 1983. Both were at Stratford for component recovery before transit to Swindon Works for eventual breaking up during 1985.

Opposite page: Ten 'Baby Deltics' were built at the Vulcan Foundry in 1959, but their single Napier Deltic 1,110hp engines proved troublesome and the entire class was recalled in 1963 for major engine rebuilds. When returned they were allocated to Finsbury Park depot and worked outer suburban passenger services from Kings Cross and over the Metropolitan 'widened lines' to Moorgate, as well as peak-time services out of Broad Street. Never having carried TOPS numbering, withdrawal of this non-standard class soon beckoned, and the first four Class '23s' succumbed during 1968, with the final two taken out of service during early 1971. One class member, D5901, was retained for use by Derby's RTC until the end of 1975, when it too was consigned for scrap at Doncaster Works. Most were sent to Stratford prior to scrapping, and D5902 was present there on 7 June 1970, having been withdrawn during the previous November. It was subsequently disposed of in September 1970 at Cohen's Cransley scrapyard, near Kettering. D5902 was sandwiched between two doomed BR Darlington Works 1952-built non-standard Class '11' 350hp shunters, 12128 and 12106, both also awaiting disposal for scrap, which was duly undertaken at Booth's Rotherham scrapyard in January 1971. (All author)

D5902

Above left: A trio of London District Class '501' slam door suburban 630v DC three-car sets, 501158, 501152 and 501188, at rest in the bay platforms at Watford Junction station on 23 April 1985. These BR Standard EMUs were built at Eastleigh Works on short 57-foot frames supplied by Ashford Carriage & Wagon Works. Designed to the same pattern as the Southern Region '4-EPB' suburban stock units with high-capacity seating and passenger-operated slam doors, and were introduced into BR service between 1955 and 1956. They operated the London Euston-Watford-Croxley Green, Broad Street-Richmond, and Broad Steet-Watford services, routinely maintained at Croxley Green depot. Earning the derisive nickname 'jail units', this referred to the bars fitted to each of the slam door windows as passenger safety measures, with some tunnels on the routes used by them having extremely tight clearances. They were withdrawn from their regular services commencing from May 1985, replaced by new automatic sliding door Class '313' units, the last units of this class having coincidentally just been taken out of service at the time of writing. Ten Class '501' DMBS carriages saw further use in departmental service as Class '97/7' battery locos, and three two-car sets were converted for use as sandite units. Two carriages were also acquired for use on the internal MoD passenger services for military personnel and their families based at the Marchwood Military Port.

Below left: '4-BUF' 'Nelson' unit 3077 arriving in the two-faced platform at Ascot on 5 December 1970, forming a Guildford to London Waterloo stopping service. One-time regulars on the SR's Portsmouth Direct 'Pompey Line' passenger services between London Waterloo and Portsmouth Harbour, and on the 'Mid-Sussex Line' via Horsham to Portsmouth, unusually this former buffet-equipped unit and three others were not immediately taken out of service when displaced by new '4-BIG' and '4-CIG' units, but were temporarily cascaded to the South Western Division's London Waterloo-Guildford-Reading via Ascot stopping services and used between 10 October 1970 and the end of January 1971, when they were finally placed into store at Wimbledon Park depot. '4-BUF' 3077, an Eastleigh 1938-built unit, was officially withdrawn from service on 13 March 1971, and after stripping of any reusable components at Selhurst depot, then followed by a period of storage at Gatwick, it was dragged to Wymondham in December 1971 for disposal on the siding used by contractor A. King & Son (*see also page 75*). The Southern Railway Bulleid-designed '4-COR' and '4-BUF' units received the nickname 'Nelsons' owing to their one-eyed appearance and daily association with Portsmouth for almost thirty-five years. (Both author)

On 9 July 1985, a mid-1985 reformed and mixed set, comprising one five-car and one two-car units with a driving trailer from '4-VEC'/Class '485' unit 045 leading, was leaving Ryde St John's Road for Shanklin. Class '03' 03079, used for shunting stock at Ryde St John's depot and workshops, and on PW trains, was stabled in the depot sidings. With the wholesale closure of all but one route on the Isle of Wight during the 1960s, modernisation of the one remaining line, the 8½-mile-long route between Ryde Pier Head and Shanklin, created several issues following the withdrawal of steam-hauled services from the end of 1966, not least its loading gauge and sharp tunnel curvature within Ryde tunnel. With DoT approval for electrification to 630v DC, the solution found was the use of refurbished London Transport (LT) 1923 Tube Stock. These 'Standard Stock' trains were formed as four-car ('4-VEC'/Class '485') and three-car ('3-TIS'/Class '486') units, and the first electric services commenced on the island from 20 March 1967. After fifty years of passenger service in London and on the Isle of Wight, where they were often subjected to the ravages of sea spray on Ryde Pier, they suffered badly from corrosion. They were replaced by more 'up to date' former LT 1938 Tube Stock, the first deliveries being made to the island from July 1989. Some 'Standard Stock' cars were cut up on the Island, but most remaining at the end of their service in the early 1990s were despatched to Bird at Long Marston and Vic Berry at Leicester. (Author)

Above left: Collision-damaged Class '33s' 33055 and 33022 at the M.&.EE. Wheel Shop Sidings at Ashford on 12 May 1991, shortly before their movement by road transporters to the M.C. Metals yard at Springburn, where they were broken up during August 1991. These two 'Cromptons' were written off between Headcorn and Pluckley on 8 November 1989, in a collision whilst hauling a Channel Tunnel construction project aggregate train to Sevington. They had slipped to a stand and a rescue loco in the form of Crewe-allocated 47455 was sent to assist. Unfortunately, the railhead conditions were so bad that upon its approach, 47455's wheels locked on when the brakes were applied, and it slid into the stricken train. Both drivers remarkably escaped significant injury: the Hither Green driver of 33055 leaped from his cab and the Ashford driver of 47455 retreated to the engine room. All three locos were condemned within days of the accident, such was the severity of the damage incurred. Some components, including the engine blocks, were salvaged at Ashford before the locos were despatched for scrap. (Author)

Below left: Class '12' 0-6-0 diesel-electric shunters 15219 and 15232 at Tonbridge MPD on 31 May 1970. Built at Ashford Works in 1949 and 1951 respectively, their SR Bulleid-design 'Firth Brown' 1930s disc driving wheels are prominent. Both were withdrawn during 1971. 15219 was cut up by Cashmore of Newport in May 1972 and 15232 at Swindon Works in February 1972. (Author)

Opposite page: Class '74s' 74009, 74003, 74005, 74010, and to the right 74007 and ex-works 33022 in the depot yard at Eastleigh open day on 30 April 1978. All Class '74s' present had been taken out of service in December 1977. Ten of these electro-diesels were rebuilt at Crewe Works between 1967 and 1968 from surplus 1958 Doncaster Works-built 650v/750v DC Class '71s'. Intended for working boat trains between London Waterloo and Weymouth and Southampton, the 650hp diesel engine allowed them to work over the non-electrified sections, obviating the need for a loco change. In practice, they proved unreliable on the Weymouth services and tended to be confined to working Southampton boat trains, where only a short section demanded diesel traction. They were also entrusted to milk trains as far as Acton yard and on overnight newspaper and postal trains out of London Waterloo in the early 1970s. All allocated to Eastleigh depot throughout their operational life, some were out-stationed at Stewarts Lane and Clapham Junction, as operations dictated. The five class members present at this open day were disposed of between August 1978 and January 1981, a mere ten years of service following their rebuild. (Author's collection)

KB

7

CAPITAL CLEAR-OUT

Opposite page: Once a commonplace sight on BR's Western Region some five years or more hitherto, by the late 1960s the few remaining former BR 0-6-0 pannier tanks, by then bearing the attractive London Transport (LT) dark maroon livery lined in black and straw yellow, were living on borrowed time. Alongside LT's 1930s-built running shed at Neasden in late 1969 were *L97* (former BR 7749 withdrawn from Bristol St Phillip's Marsh MPD in September 1962), *L91* (former BR 5757 withdrawn from Westbury MPD in December 1960), and *L98* (former BR 7739 withdrawn from Neath MPD in November 1962), all set aside for disposal, which took place at Neasden during 1970 by Sheffield contractor Steel Breaking & Dismantling Ltd. LT had purchased thirteen BR pannier tanks, which saw service variously between 1957 and 1971, with two earlier acquisitions exchanged at Swindon Works for two in better condition. LT steam traction was eliminated from the engineer's train duties originating from Neasden and Lillie Bridge depots by June 1971, so ending the last regular main line steam diagrams in the Capital, almost four years after Southern Region's steam finale at Waterloo station in July 1967. (Author's collection)

Above right: LT ballast motor car *L141*, a 1975 rebuild by Acton Works from Metropolitan-Cammell 1938-design Tube Stock, formerly driving motor car 11067. Photographed at BR's Fratton depot sidings on 16 February 1991, it was pending disposal on site by a local scrap contractor. These motor cars, some latterly presented in the standard LT Engineering Department's yellow livery, usually operated in pairs, and were deployed on engineering diagrams hauling ballast wagons. Their interiors were ideal for conveying track maintenance personnel and small tools and equipment to designated work sites during overnight possessions on the London Underground network, and some were additionally fitted with weed-killing equipment. (Roy Burt)

Below right: Four further ballast motor cars were also present at Fratton on 16 February 1991. In the line-up awaiting breaking up were *L145* and *L144*, both rebuilt at Acton Works in 1975 from 1938 stock driving motor cars 11027 and 10257, respectively, and *L152* and *L153*, rebuilt at Acton Works in 1978 from driving motor cars 10266 and 11266, respectively. All had been rebuilt as ballast motor cars following the cascading of LT's 1972 underground stock to the Bakerloo Line, superseding the last of the 1938 stock then remaining in service. They were normally used for engineering work on the Underground network, operating in pairs in top and tail formation, although a battery loco and ballast motor car formation was also employed whenever the conductor rails were de-energised. (Roy Burt)

Above left: In 1969, during the last full year of steam workings on the 'Watford Tip run', LT pannier tank *L95* (ex-BR WR 5764 withdrawn in May 1960 from Old Oak Common MPD) was photographed at Croxley Tip shunting open bogie wagons containing waste material from the Neasden power station demolition. These regular weekday steam diagrams continued into 1970, until substituted by battery locos. The Neasden-based pannier tank with laden ballast wagons marshalled between two brake vans used the Metropolitan line via Croxley, and would then round its train at Watford, taking on water before heading for the tip sidings, where access was gained from the up line, just before Watford South Junction. When finally taken out of service in 1971, *L95* was sold to the Severn Valley Railway for preservation. The LT brake van *B.582* was one of a fleet of six built in 1962 at BR's Ashford Carriage & Wagon Works. (Author's collection)

Below left: The last LT engineer's train diagram for steam haulage unceremoniously occurred during the night of 14–15 May 1971, when pannier tank *L90* was allocated to the regular Lillie Bridge to Neasden, via Rayner's Lane, ballast train. All that remained thereafter was for the public to officially bid a farewell to the enigmatic 'red panniers'. A 'Farewell to Steam on the Underground' special event was staged by LT on Sunday 6 June 1971, when pannier tank *L94* (formerly BR 7752 taken out of service at Tondu MPD in December 1959), was entrusted to a commemorative steam working from Moorgate to Neasden depot, thrown open to the public. The platforms were thronging with well-wishers as it arrived with its demonstration engineer's train on the non-electrified Metropolitan 'Widened Lines' at Barbican station. Following the event, *L94* passed into preservation at Tyseley Locomotive Works. (Author)

Opposite page: Midland Railway (MR) 1884-built Johnson '1F' 0-6-0 tank 41734 of Staveley (Barrow Hill) MPD acting as works pilot at Staveley Ironworks in 1964. Its longevity was assured by a legally binding 1866-dated document signed by the MR and known as the '100 years agreement', obliging it and any successors to supply shunting locos to the Staveley Coal & Iron Company. 41734 had been at Staveley since July 1958, and remarkably remained in BR service until December 1966, latterly working from Langwith Junction MPD. It was cut up during April 1967 by Arnott & Young at Rotherham. Shed mate June 1880-built 41708 was also taken out of service at Langwith Junction in December 1966. It was moved initially to Leeds Holbeck depot before passing into preservation, today appropriately on display in the listed and renovated Barrow Hill roundhouse. (A.E. Durrant – author's collection)

8

DECLINE OF INDUSTRY

Above left: Two locos set aside for disposal at Mardy colliery on 19 August 1970, pannier tank *Mardy No.4* (ex-BR 9792 withdrawn from Neath MPD in March 1964) and Newcastle-built Robert Stephenson & Hawthorns 0-6-0 saddle tank 125 (works No.7099 built in 1943). 9792 was scrapped by a Newport-based contractor in September 1973. Both locos had proved unsuitable for the arduous duties demanded at this remote colliery location at the head of the Rhondda Valley. (Roy Burt)

Below left: Pannier tank 7754 (withdrawn from BR service at Wellington MPD in January 1959) was one of the last two active NCB fleet members of the once numerous GWR '5700' Class built between 1929 and 1949. On 5 May 1975, it was pending attention inside Mountain Ash shed with Andrew Barclay 0-6-0 saddle tank *Llantanam Abbey* (Works No.2074 built in 1939). Arriving from Talywain colliery in 1970, the North British 1930-built loco was unpopular at Mountain Ash, considered entirely unsuitable for its undulating trackwork, and consequently it saw limited service there compared to its thoroughbred industrial counterparts. Upon closure of the Mountain Ash system, it was donated by the NCB to the National Museum of Wales, who in turn placed it on permanent loan to the Llangollen Railway. Its poor mechanical condition following neglect during NCB service necessitated protracted restoration, which was nearing completion during 2023, over four decades since withdrawal from industrial service. (Author)

Opposite page: Situated on the clifftop overlooking the Solway Firth, the seaward view from NCB's Ladysmith Preparation Plant in Cumberland could certainly be breathtaking on a clear day, but at times it could also be a bleak and foreboding place. The locos were worked relentlessly hard by the enthusiastic crews, with traffic requirements demanding minimal time for sprucing them up. Although these dumped locos had seen little, if any, work on the interconnecting Ladysmith and Haig colliery system at Whitehaven, they had been rain-washed to reveal their original faded livery, in the process showing just how the salt-laden air and accumulation of char in the smokebox had played a part in their overall external deterioration. Languishing on the 'scrap line' on 15 August 1969 were 0-4-0 saddle tanks *Risehow No.2* (Yorkshire Engine Co. Works No.2432 of 1948), *Spider* (Hawthorn Leslie 2623 of 1905), and *Risehow No.1* (Yorkshire Engine Co. 2430 of 1948). Austerity 0-6-0 saddle tank *Revenge* (Hunslet 3699 of 1950) was busying itself beyond at the washery, positioning BR 21-ton hopper wagons, a regular duty it would continue to undertake until May 1973, after which it also was to share the fate of those already discarded at Ladysmith. (Roy Burt)

Above: On a murky 24 August 1971, Robert Stephenson & Hawthorns 0-4-0 saddle tank *Stella No.2* (Works No.7799 built in 1954) gradually being dismembered at Dunston's Norwood Coking Plant, with shunting at the plant already having been given over to diesel traction in 1969. (Author)

Opposite page: The last days of the Ashington Collieries system – on 17 September 1986, BR Swindon Works-built Class '14' D9525, running as NCB 507, clagging well as it worked past the Ashington colliery headstocks with a long rake of internal hopper wagons. Andrew Barclay 0-6-0 diesel-hydraulic (Works No.488 built in 1964) was also gainfully employed with a rake of loaded internal-user wagons. The ground shunter clearly had a busy day ahead, judging by the number of wagons present during the final days of this colliery and its railway. D9525 was purchased for preservation, but Ashington colliery was to close during the following month. (Roy Burt)

Above left: A remarkable line-up of redundant British Steel Corporation locos alongside the River Tees near Teesport on 28 October 1977, including two unique Thomas Hill articulated 'tridem' locos, as well as conventional Yorkshire Engine and Rolls-Royce 'Sentinel' locos. The 1969-built Thomas Hill articulated diesel-hydraulic units (Thomas Hill Vanguard Works Nos in the range 201V–211V) each had a Rolls-Royce 'Type C6TFL power unit delivering 278hp. The 'master' driving unit could attach a 'slave' power unit at each end, thereby becoming an 834hp unit. One such 'master and slave' unit is seen at the head of this row, followed by a spare 'slave unit'.

Below left: The shell of former BR Percy Main-allocated Class '04' Drewry D2326 in the yard at NCB Manvers Main Workshops on 3 August 1971. Withdrawn from BR service in August 1968, it had been cannibalised for spares to keep other sister Drewry locos in service at several Barnsley area collieries. D2326 was cut up during February 1976. Perfectly aligned and glimpsed through the engine compartment of the Drewry was Hunslet 'Austerity' 0-6-0 saddle tank (Works No.3701 built in 1950), also out of service.

Opposite page: The railways serving the NCB Wheldale and Fryston collieries were well known in industrial railway circles for the remarkable 'Austerity' saddle tank underfeed stoker trials that were conducted by the Hunslet Engine Company of Leeds between 1981 and 1982. However, this single emotive scene perhaps epitomises the wholesale decimation of the British coal industry during the 1980s, and BR's consequent reduction in coal traffic and contraction of its fleet resources. This was Wheldale colliery on 20 September 1986, one year before its final closure, but by this time already appearing forsaken. Some of the last remaining diesel shunting locos there were in the process of being despatched for scrap, and in the foreground was 'Sentinel' four-wheel diesel-hydraulic *N.C.B. 7* (Works No.10183 built in 1964), transferred to Wheldale from Lofthouse colliery in January 1973. Apart from tyre turning at BR's Doncaster depot in June 1981, it had held out at Wheldale until December 1986, when it was finally broken up by Sheffield-based contractor Marple & Gillott Ltd, the few reusable components having already been stripped by Hunslet. The Hunslet 0-6-0 diesel-hydraulic behind (Works No.7352 built in 1973) had plied the tracks at Wheldale since new, but its later construction in comparison with other NCB locos which managed to survive up until those depressing times in the coal industry was no guarantee of it avoiding the cutter's torch, a mission duly conducted at the Hunslet works in Leeds shortly following this, once any potentially reusable components had been recovered. (All Roy Burt)

9

COMMERCIAL BREAKERS

With the BR workshops unable to cope with the volume of steam locos requiring disposal during the late 1950s and early 1960s, private contractors undertook the scrapping of hundreds of steam locos, some at BR depot sites, but primarily on their own private premises.

Above: 34017 *Ilfracombe*, withdrawn from service in October 1966, awaiting its fate at Buttigieg's yard at Newport in June 1967, having arrived in March 1967. It was the final Bulleid Pacific to be cut up, and its longevity was undoubtedly on account of it being constantly hemmed in at the end of this siding by other new arrivals, but its turn eventually came in December 1968, some two years after its official withdrawal date at Nine Elms MPD. (Charlie Cross)

Some of the major players in this field, although not by any means an exhaustive list of scrap merchants, were T.W. Ward, John Cashmore, George Cohen, Bird Group, Albert Draper, J. Buttigieg, Arnott & Young, A. King & Son, and several others referred to in the chapters of this book, of course not overlooking Woodham Bros of Barry. The demand for ferrous and non-ferrous metals was an integral part of Britain's industrial economy, feeding the voracious demand of the steel industry of that time, thus ensuring that when redundant locos came up for tender by both BR and the larger private industry users of locos, they were eagerly snapped up, followed in most cases by their rapid demise, and BR's loss proved to be the scrap merchants' gain. BR's decision to outsource disposal of some but not all locos and rolling stock did not cease with the wholesale clear-out of steam traction by 1968, for many of the early pilot scheme diesel locos, multiple units, and other life-expired locos and rolling stock also followed the same way, although players in the field gradually changed to Vic Berry, Marple & Gillott, M.C. Metals, and C.F. Booth, to name but a few, becoming prolific from the 1970s and 1980s as more and more locos and rolling stock became available, especially those with blue asbestos lagging requiring specialist treatment.

Opposite page below right: In July 1967, LMS Class '3F' 0-6-0 tank No.47305 stood dejected at Troon Harbour, in the company of a Stanier 'Black Five' 4-6-0, awaiting the cutter's torch. The 'Jinty' had worked from Barrow depot during its final active year, and was well accustomed to the ravages of the salt-laden sea air. Withdrawn during February 1965, and following six months of storage at Barrow, it was towed to Arnott & Young (West of Scotland Shipbreakers) at Troon Harbour in the September of 1965, and was not to meet its end until April 1968, such was the volume of locos requiring processing. The obsolescent cattle wagons also for disposal, just visible beyond the palisade fencing, are noteworthy. Arnott & Young had several scrapyards in Scotland, most of them strategically centred around the Central Belt, close to its many steel plants at that time.

Above right: With its dome askew and in a deplorable condition overall, former Guildford depot's 'N' Class 2-6-0 31858 was standing in the company of un-rebuilt Bulleid 'West Country' Pacific 34033 *Chard* and a BR Standard '3MT' 2-6-4 tank at Buttigieg's Newport scrapyard in June 1966. Both had been withdrawn in December 1965, and despatched to this South Wales scrapyard from Eastleigh depot during March 1966. 31858 possessed the distinction of having worked the 11:05 Reading-Redhill diagram on the last day of steam-hauled passenger services on this route, on 2 January 1965. These locos were still awaiting their fate during mid-September of that year, such was the frequency of loco arrivals and the volume of processing work required at that time, particularly with those originating from SR depots. Nothing went to waste, judging by the small stack of coal in the foreground, recovered from what little remained in the tenders.

Below right: A depressing scene at Buttigieg's yard in June 1967, especially for lovers of Bulleid Pacifics, where 'West Country' 34032 *Camelford* and other classmates were awaiting disposal. 34032 was withdrawn from Nine Elms MPD on 2 October 1966, but was temporarily reinstated for a brief period in December 1966, and was again placed into store at Nine Elms in February 1967 prior to its removal to Newport shortly thereafter. It ended its days in the company of classmates 34005 *Barnstaple* and 34009 *Lyme Regis*, with 34017 *Ilfracombe*, its tender, just visible to the right. 34032 was dealt with during September, 34005 during October, and 34009 was reduced to scrap metal by the November of that year. A total of twenty-four Bulleid Pacifics (six 'Merchant Navy' and eighteen of the 'West County'/'Battle of Britain' Class) were dealt with by Buttigieg at Newport. (All Charlie Cross)

Right: The skeletal remains of Stanier '8F' 2-8-0 48290 in an advanced stage of disposal at Buttigieg's yard in June 1966. It had been withdrawn from Sutton Oak depot during August 1965. The former War Department loco No.70438, built by Beyer Peacock in 1941, had seen service in Persia. It was repatriated to the UK circa 1949 and was sold on to BR, initially allocated to Crewe South in September 1949, never receiving a pre-nationalisation running number. (Charlie Cross)

Left: As a location previously referred to in several captions in Chapter 1, this was Albert Draper's scrapyard on 7 April 1968, with Fairburn '4MT' 2-6-4 tank 42251, withdrawn from Low Moor MPD in September 1967. Also, in the line-up for disposal were 'WD' Class 2-8-0s 90605 and 90721, both taken out of service from Normanton depot in September 1967. (Author)

Above right: To the north side of Queenborough station on the Isle of Sheppey was a branch line, 1½ miles in length, running alongside the banks of the River Swale to a wharf on Chetney Marshes. This was where over twenty condemned steam locos, primarily from Feltham, Eastleigh, and Nine Elms depots, were scrapped during the mid-1960s. These included four 'M7' 0-4-4 tanks, three 'Q' Class 0-6-0s, and Bulleid Pacifics 34003 *Plymouth*, 34029 *Lundy*, 34061 *73 Squadron*, and 34091 *Weymouth*. The wharf was owned by Settle Speakman & Co. Ltd and between 1908 and 1962 was used for shipments of coal from Scotland. It was subsequently leased to Lacmots Ltd (from the mid-1960s Shipbreaking Queenborough Ltd) for the disposal of BR locos, rolling stock, and redundant trackwork, as well as for ship-breaking activities. After ship-breaking had ceased in the 1970s, a rolling mill was established on Settle Speakman-owned land at Queenborough, and the branch was subsequently used for moving imported steel billets to the mill. The Queenborough Rolling Mill Co. Ltd had Italian connections, and a small loco workshop was established there. From around August 1971, redundant BR and industrial locos were purchased, thought to be around twenty-five ex-BR locos in all, mainly of Classes '03' and '04', with at least half of them being exported to Italy via Sheerness Docks, although the precise number has not been verified. Some former BR and industrial locos were retained at Queenborough, for working the exchange sidings, in the rolling mill, and along the branch to the wharf. A visit to their depot yard on 10 May 1990 revealed former BR Class '03' D2070 out of use. Built at Doncaster Works in 1959, and initially delivered to West Hartlepool depot, it remained around Teesside working from West Hartlepool and Thornaby depots until its withdrawal and purchase in June 1972. After several years of gainful employment at Queenborough, it was sold on in 1990 to the Churnet Valley Railway, but circa 2001 it was eventually sold for scrap, by then in the ownership of Cotswold Rail at Gloucester. (Author)

Below right: On 21 August 1976, former BR Class '08' D3099 and Class '03' 03164 were observed by the main entrance to the mill's Rushenden Road depot at Queenborough. 03164 was subsequently exported to Italy in March 1977, but the Derby Works 1955-built D3099 is believed to have been cut up on site by 1980. Imported steel billet traffic had ceased at Queenborough by April 2010, when the wharf branch was finally closed. The rolling mill, by then under the ownership of Invicta Merchant Bar plc, was subsequently closed in February 2013. (Roy Burt)

Observing the more stringent legislation then being introduced appertaining to the disposal of railway vehicles containing blue asbestos material, the scrapyard of A. King & Sons (Metal Merchants) Ltd was established in 1973 on the site of a yard at Snailwell, previously owned by W.E. Knappet & Sons Ltd. Until a suitable location could be obtained and developed, the company had used an isolated section of closed railway at Wymondham for such activities (*as previously illustrated on page 75*). The remote Snailwell site was located on the north side of the Ely–Newmarket line. Whitemoor Yard at March and the down sidings at Ely became strategic staging posts for BR stock awaiting eventual disposal in the Snailwell incinerator. In the low winter sunshine at Snailwell on 17 January 1983, a line of Metropolitan-Cammell Pullman coaches, with Pullman Kitchen Second (PSK) *Car No.338* in the front row, await possible further use or disposal, the latter eventually taking place for this vehicle. Alongside, a Cravens Class '105/2' MBS unit, Sc51480, was awaiting entry into the incinerator. The PSKs were made redundant from the ECML services between 1968 and 1971, replaced by air-conditioned BR Mk2d coaches. Some PSK Pullmans eventually found alternative use as licensed bars or ancillary hotel restaurant facilities, although most have since been disposed of. Yard shunter, Andrew Barclay D2956 (*see page 84*) can be seen in the left distance. (Roy Burt)

Above left: BR Swindon Class '126/2' Inter-City MBS Sc51040 was awaiting processing at Snailwell on 17 January 1983. The business was taken over by Mayer Newman & Co. Ltd in 1983, and subsequently by Mayer Parry Ltd in February 1991.

Above right and below right: On 17 January 1983, former BR Class '03' 03012 was attached to Cravens MBS unit M50778, ready to place it into the specialist blue asbestos incinerator at Snailwell. The company was one of only a small number of specialist contractors licensed to process former BR vehicles containing this hazardous material. The first locos to arrive on the site when operations began in earnest were Class '03s', 03012 from March TMD and 03020 from Norwich TMD, both during December 1975, followed by '01' Class D2956 transferred from King's Norwich yard in 1981, and 03180 from BR Norwich TMD in July 1984. Other industrial locos for light shunting duties had also been used at various times. (All Roy Burt)

Above left: Founded in 1920, metal trader C.F. Booth was associated with the dismantling of buses, locos, main line, and underground carriages and wagons at several Yorkshire sites, but the main facility for railway vehicles was that established in the 1960s at the Clarence Metal Works in Rotherham, dealing in ferrous and non-ferrous metals. Since then, and at the time of writing, many condemned main line and industrial locos and vehicles continue to meet their fate at this busy scrapyard. The 1980s was an especially busy time, the yard processing many locos following the mass closure of collieries in the Midlands, Yorkshire, and the North East, also coinciding with a period of severe rationalisation of the BR diesel and electric loco fleet. The most controversial was the dismantling there of most of the ill-fated APT-P vehicles in 1987. On 12 February 1993, Class '08' 08334 was awaiting its turn for disposal, and already being dealt with were 47456 and 31425. One of the yard's shunters on this visit was former Shirebrook colliery's Hudswell Clarke 0-4-0 diesel-hydraulic (Works No.D1344 built in 1965). Booth's made short-term use of some arrivals before they also followed in the same way as their earlier counterparts, although some more fortunate examples were sold on for further industrial use, or occasionally into preservation. (Author)

Below left: English Electric/Baguley 1959-built *N.C.B. 15* from the closed 1,500v DC Harton Electric Railway at Booth's scrapyard on 24 January 1990, in the company of other redundant NCB locos. The low-height cab enabled access through Factory Tunnel, between St Hilda's yard serving Westoe colliery, and the staiths. The installation of a rapid loading system for MGR trains at Westoe saw the last coal shipped in July 1988. The electrified railway, latterly used only for spoil trains, was closed in July 1989. Apart from the pantograph positioning, the locos bore a remarkable resemblance to the NER Darlington-built Bo-Bo 1,500v DC locos (26502–26511) which worked the freight route between Shildon and Newport (Co. Durham) between 1915 and 1935. (Author)

Opposite page: A Booth's ground shunter grips his shunting pole and observes a mixed rake of vehicles being positioned into place on 18 November 1988. Although around the time of *Speedlink* becoming one of BR's new operating sectors designated *Railfreight Distribution*, this change did not halt the cull of freight rolling stock. A limestone 'CBA' covered hopper wagon was in this consignment for disposal. Based on the MGR coal carrying hoppers of the time, BR Shildon Works built fifty-two such wagons in 1969 for limestone traffic to steelworks, however the contraction of the steel industry saw them being despatched for scrap in the late 1980s. (Roy Burt)

Poclain
125
S CO. LTD.
Railfreight

Above left: George Cohen developed a scrapyard at Cransley from around 1960, located on the closed Cransley Ironworks site, adjoining the former Kettering–Loddington ironstone branch line. Initially processing BR and LT rolling stock, the first steam locos arrived in March 1964 from the SR, but as BR disposals increased during the 1960s, an increasing number originated from the Southern, Western, Midlands and Yorkshire area depots, and latterly from the North West. In all, around 250 BR steam locos and a handful of industrial saddle tanks from the local ironstone industry had been scrapped at Cransley by 1969. Classes had included SR 'S15s', 'Schools', 'Ws', and 'Q1s', GWR 'Halls', 'Granges', and 'pannier tanks', LNER 'B1s and ROD '04s', LMS '4Fs', 'Jinties', '8Fs', 'Black Fives', Ivatt '2MTs' and '4MTs', and BR Standard '3MTs' and '9Fs', among others. By the late 1960s, Cohen's attention had already turned to the disposal of BR diesel traction, both locos and multiple units, as well as an occasional industrial diesel. Many arrivals had originated from Stratford depot including 'Baby Deltics' and examples of Classes '15' and '16'. Additionally, 'Hymeks' and 240hp, and 350hp shunting locos were similarly dealt with at Cransley, all amounting to a phenomenal variety having passed through the facility over a period of around two decades, up until its closure and site clearance in September 1981. Still in original BR green livery, 'Derby Lightweight' MBS E79025 and DTC E79668 awaited their fate at Cransley on 21 June 1970. In the background was MBS E79184, all a victim of swingeing line closures of the 1960s.

Below left: Following withdrawal by BR, Swindon Works-built Class '03' D2176 found short-term use as a yard pilot at Cransley. Looking the worse for wear on 21 June 1970, it was eventually scrapped during November 1971. Accepted into BR capital stock at Ashford MPD in December 1961, its last BR task was shunting Crewe Works until its withdrawal in May 1968. The *door to door* intermodal containers are noteworthy, including one carrying the *Pedigree Prams* company name, these containers having been conveyed on trailers from many BR freight depots for delivery by road, hauled by the once ubiquitous Scammell Scarab three-wheeled mechanical horses.

Opposite page: BR blue 'Derby Lightweight' MBS M79184 was anticipating the imminent torch at Cransley on 21 June 1970. A victim of the numerous branch lines axed in the 1960s, the non-standard 'Yellow Diamond' coupling code was guaranteed as a candidate for early disposal. All non-blue square coupling code DMUs were disposed of as soon as they could be released from traffic, replaced by 'cascaded' blue square-coded DMUs made surplus either by line closures, service reductions, or route electrification. (All author)

Above left: Two unidentified and partly dismembered 'Jinty' Class '3F' 0-6-0 tanks at Cohen's scrapyard in early 1967, during the period when sixteen of the class from Wolverton, Crewe, Stoke, Chester and Holyhead were dealt with at Cransley. (Author's collection)

Below left: The Garnham, Harris & Elton scrapyard at Chesterfield in February 1968 contained BR Standard '4MT' 75055 and BR Standard '3MT' 2-6-0 77002 next in line for the torch, withdrawn from York and Stoke MPDs respectively in June 1967. In the foreground, Peppercorn 'K1' 2-6-0 62048 had already been partially dismantled, having arrived from West Hartlepool MPD in August 1967. (Graham Wignall – author's collection)

Opposite page above right: One of the more unusual BR locos dealt with at Cransley was Ruston & Hornsby Class '48DS' departmental *DS1169*, noted there off the track in a pile of scrap on 23 June 1973. It was delivered new in 1946 to the Bristol Aeroplane Company at Weston-super-Mare, but was subsequently sold on to BR in 1948 for use at the Folkestone Warren P.W. depot. In 1962 it was transferred to Broad Clyst Engineering depot, and in 1965 to Yeovil Junction engineer's sidings, until its final withdrawal came in October 1972. (Roy Burt)

Right: In October 1987, a stack of BR Type 2 Class '25' and '27' loco bodies were awaiting their fate at Vic Berry's Leicester scrapyard. Nearest the camera were ex-BR Scottish Region 27049 (previously D5388 and 27105), 27053 (previously D5397 and 27109) and 27037 (previously D5389). This Leicester scrapyard was established in 1973 on the site of the former GCR Braunstone Gate goods yard, adjoining the Marylebone–Leicester line south of the former Leicester Central G.C. station. The yard initially broke up BR freight and passenger vehicles, until the first BR locos arrived there in 1983. Over the ensuing eight years a varied selection of diesel and electric classes were processed, in addition to a continual flow of various DMU and EMU classes, and BR Mk1 passenger vehicles, as well as a handful of CIE locos from the Irish Republic. Vic Berry acquired locos from around the entire BR network, with representatives of Classes '02', '03', '08', '20', '25', '26', '27', '31', '33', '37','40', '45', '47', '50', '76', '82', '83', '84', '85', and '97' processed. The facility specialised in the prolonged process of blue asbestos removal through incineration, which led to a backlog of railway vehicles awaiting disposal, and a viable solution proved to be firstly removing major internal components, then the bogies, battery boxes, fuel tanks and buffers, thus enabling the loco bodies to be stacked three-high. A similar space-saving solution was also adopted for coaching stock. There was no rotational system applied to the stacks, which were gradually removed as scrapping capacity dictated. Locos were also disposed of by Vic Berry's roving operatives at several BR sites. An earlier than envisaged end came to the Leicester operation when a severe fire took hold at the yard during the early hours of 10 March 1991, leading to the airborne dispersal of asbestos material across the city. Vic Berry was ordered to immediately clear the site, but it was left seriously contaminated and, resultantly, the company ceased trading in June 1991. It was eventually redeveloped as *Bede Island*, under the auspices of the late 1990s *Leicester City Challenge Project*. (Author's collection)

A general view of Vic Berry's Leicester scrapyard at 6.30 a.m. on the murky Sunday morning of 14 June 1987, with many newly arrived Class '25's' awaiting stripping and stacking for eventual future 'torching'. (Author)

The dismembered remains of Class '47/4' 47415, and the smouldering hulks of former 5ft 3in gauge Metropolitan-Vickers of Manchester 1956-built CIE Class '201' locos 214 and 215, being processed at Vic Berry's scrapyard on 29 June 1990. (Author)

An unusual comparison of locos awaiting disposal at Vic Berry's yard on 19 August 1990. Class '85' AC electric locos 85019 and 85025 stand alongside ex-CIE Class '201' Metro-Vick diesels 208 and 219, both transported across from the Irish Republic by road in a convenient logistical swap with coaching stock requiring return transportation to the Irish Republic. (Author)

Above: The 'VEA' ventilated van was a traditional wagon design, and 550 were refurbished and air brake equipped in the early 1980s for use on BR's 'Speedlink' trains and ad-hoc military traffic, including the conveyance of explosives. A decline in 'wagon-load' freight by the early 1990s led to a mere handful seeing further use, in BR departmental service, but most were consigned for scrap. At T.J. Thomson's Millfield scrapyard at Stockton-on-Tees in July 1992, over fifty such vans were awaiting disposal. (Author)

Right: The end of the line for two LMS/BR-designed late 1950s-built standard vacuum-fitted 12-ton ventilated vans at Bird's Cardiff scrapyard on 20 August 1970, with the wood plank or plywood sides and body panels being incinerated prior to the skeletal remains being torched. The white-painted corrugated end panels could suggest these vans to have been insulated for meat traffic conveyance, but the destruction of the side panels precluded precise identification of their former function. (Roy Burt)

Above left: At the Bird Group's Long Marston yard on 22 August 1988 was former BR 1960-built Yorkshire Engine Class '02' D2857 looking much the worse for wear after having suffered an engine fire. It was paired with a former GWR 'Iron Mink' unfitted 10-ton S&T tool van, DW80888. During its short ten-year career with BR, D2857 had been initially based at Bank Hall MPD and worked around Merseyside on duties such the station pilot at Liverpool Exchange and shunting around the Liverpool docks lines. It was reallocated to Speke Junction MPD in 1966, and nearby Allerton TMD in 1968, by which time the requirement for lighter four-coupled shunting locos had diminished. BR disposed of it soon after its withdrawal in April 1971, when it was purchased by the Bird Group, finding work as the Long Marston yard shunter for a further two decades, until June 1992, when it too went the same way as others before it.

Below left: On 30 March 1970, the Long Marston scrapyard of Bird's Commercial Motors (later Bird Group) contained diminutive Hunslet 0-4-0 diesel-mechanical locos *Gert* (Works No.1847 built in 1936) and *Daisy* (Works No.1898 built in 1939) sandwiched between two larger main line cousins, 1961-built Class '03s', D.2188 taken out of service at Swindon MPD in May 1968, and ex-Landore TMD's D.2146, taken out of service in September 1968. Both had experienced outrageously short working lives with BR. Also in the row on the left was ex-Stratford TMD's Drewry Class '04' D2215 (formerly 11121), which had been withdrawn in March 1969, but had at least seen almost fourteen years of service following delivery into BR capital stock in July 1955. The 40/44hp Hunslet industrial locos had been supplied new to the MoD's Royal Ordnance Factory at Kings Meadow near Nottingham. Although the Drewry and the two Hunslets were almost immediately disposed of soon after arrival at the scrapyard, the two '03s' remarkably survived at Long Marston until 1978, but became increasingly derelict, having been cannibalised for spares over the ensuing eight years, coming to the aid of those of the class soldiering on in industry. During October 1971, D.2146 was at least found limited use when it was loaned to the adjacent Royal Engineer's Park for a programme of Army re-railing training exercises. Over many years, Bird's scrapyard had disposed of commercial lorries and buses and a variety of BR, LT, and industrial locos and rolling stock. The first steam locos for disposal arrived at Long Marston in June 1965, followed by BR and industrial diesels from around 1967. The last loco disposed there was in November 1994. (Both author)

Left: Former MoD Army Department Andrew Barclay 0-8-0 diesel-hydraulics 624 *Royal Pioneer* (Works No.504), 622 *Greensleeves* (Works No.502), 620 *Sapper* (Works No.500), and 623 *Storeman* (Works No.503) at Marple & Gillott's scrapyard at Attercliffe, Sheffield, during the early morning of a wet 4 June 1985. All nameplates had been removed and other identification painted over. Six of these powerful 61-ton 310hp locos had been supplied to the Army, and they had been employed on the MoD Bicester and Kineton Military Railways. Unusually, they had been delivered to the military establishments by rail from Kilmarnock between 1964 and 1966. During early 1985, they had become surplus to requirements and were immediately disposed of, their long wheelbase not being to their advantage for finding further use elsewhere in industry. (Roy Burt)

Right: Former Manchester-Bury BR standard design 1,200v DC side contact third rail Class '504' Driving Trailer Open Second (DTOS) M77164 was reformed with Class '302' unit 244 for working on the Fenchurch Street-Shoeburyness 'misery line' services, replacing one of the Class '302' unit's accident-damaged vehicles. It was re-prefixed E77164, and this 'hybrid' formation ran in service until early 1985, when another Class '302' driving trailer became available. E77164 was then sent for scrap, and it was photographed in the Attercliffe scrapyard of Marple & Gillott on 24 April 1985. Of the twenty-six units, which were originally built at Wolverton Works in 1959 and maintained in the 1856-built Grade II listed building at Baron Street and now used by the heritage ELR, one two-car unit survives, formed of M77152-M65451, which is now in the care of the 'Class 504 Preservation Society' based at the ELR. The restoration of the driving trailer M77172 was progressing well in 2023, with the matching DTOS waiting next in line, but requiring extensive restoration work by volunteers. (Author)

Above: M.C. Metals' scrapyard on 23 March 1986, with Vulcan Foundry 1946-built 0-4-0 diesel-mechanical yard shunter *Army 227*, a standard loco supplied to various military depots countrywide during the early post-war years. This Glasgow yard, conveniently located adjacent to BR's St Rollox Works, handled many BR diesel and electric locos, and a line of Class '08' shunters, 08446, 08564, and 08433, were awaiting processing on this visit. Originally stationed at the Royal Army Ordnance Corps' base at Stirling, it was subsequently purchased by Direct Rail Services, earmarked for use as the Kingmoor depot's pilot loco, but its 153hp power output proved inadequate. It is now at the Eden Valley Railway, Warcop. (Author)

Left: The remains of an unidentified '48DS' Class Ruston & Hornsby four-wheel diesel-mechanical loco beyond redemption at Marple & Gillott's Attercliffe scrapyard on 25 November 1983. Over 200 such locos were built between 1937 and 1967 and they proved popular for light shunting duties in industry, although BR acquired just two, used in departmental service, *ED10* and *DS1169* (*see page 114*). (Roy Burt)

10

HANGING IN THE BALANCE

Right: Class '73/1' 'Electro-diesel' 73134 *Woking Homes 1885-1985* was a mere bodyshell at Wabtec's Falcon Works, Loughborough, by 2017. Originally emerging from the Vulcan Foundry in August 1966 and numbered E6041, it served its early years working alongside the last vestiges of steam traction, and the class proved to be an ideal replacement for the Southern Region's steam forebears, used on 'pick-up' freight duties. They also proved ideal for shunting the local freight facilities in the remaining non-electrified station yards. Named after the Southern Railwaymen's convalescent home in Woking, 73134 was acquired by English, Welsh & Scottish Railway (EWS) after privatisation, but would be taken out of service in 1999, upon EWS' fleet rationalisation. Purchased for preservation, it spent time on the heritage Dartmoor and Weardale Railways, then was found use as a site shunter at Staveley (Barrow Hill) roundhouse. Eventually placed into store at Nemesis Rails' Burton depot, it was purchased by GB Railfreight, earmarked as a potential spare body for the company's fleet of Class '73/9' locos then being rebuilt at the Falcon Works for use on the Caledonian Sleeper services, but it was never required. Following clearance of the Brush Wabtec-owned works, 73134 was despatched to Booth's Rotherham scrapyard for disposal and was duly reduced to scrap metal in mid-September 2022. The first six 'JA' prototype batch of electro-diesel locos were introduced onto the SR in 1962, numbered E6001–E6006 (later Class '73/0' 73001–73006). So successful was the design that a further order was placed with English Electric, and an additional forty-three production locos, E6007–E6049 (later Class '73/1' 73101–73142), were built at the Vulcan Foundry. Classified as 'JB', they were fitted with an improved traction motor and bogies, allowing increased tractive effort and a top speed of 90mph. All were built to the narrower 'Hastings Gauge', allowing greater flexibility in their deployment. They drew power from the 650/750v DC third-rail through eight collector shoes mounted on their bogies and could operate on non-electrified lines using the on-board 600hp diesel engine. Considered versatile mixed-traffic locos, they were later used on the 'Gatwick Express' passenger services, fourteen being allocated to these duties, the remaining class members passing to the freight operator EWS following privatisation. Remaining loyal to SR metals, EWS withdrew most class members during the late 1990s, with several being sold on to other operating companies, or passing into preservation. (Author)

Above left: A line of nine AC electric locos stored unserviceable at Longsight depot on 8 January 1983, pending a decision on their future. In the row, in order, were 82004 (E3050), 83007 (E3030), 83009 (E3032), 83010 (E3033), 82008 (E3054), 83005 (E3028), 83013 (E3098), 83008 (E3031), and 83002 (E3025). Five had previously experienced an extended period of storage at the former Bury MPD in the early 1970s (*see page 127*). A decision was eventually taken during 1983 to dispose of most of them: seven were consigned to Vic Berry's Leicester yard and cut up during late 1984, but two enjoyed reinstatement for empty coaching stock (ecs) transfers between Stonebridge Park carriage sidings and London Euston station. 82008 remained in service until December 1987, and 83009 was eventually taken out of service in March 1989. (Roy Burt)

Below left: AC electrics 83012 (E3035) and 82008 (E3054) stabled at London Euston on 25 October 1983. Built by English Electric at the Vulcan Foundry in the early 1960s, the last Class '83' locos to remain in service, 83009 (E3032), 83012 (E3035) and 83015 (E3100), were retained for use on the ecs workings, along with Beyer Peacock-built Class '82s' 82005 (E3051) and 82008 (E3054). Delivered new to BR in July 1961 as 'AL3' Class E3035, final withdrawal for 83012 came in 1988, but it happily survived into preservation, the sole member of the class to do so, and it is currently at Barrow Hill in the care of the 'AC Locomotive Group'. 82008, delivered new to BR in November 1961 as 'AL2' Class E3054, was also fortunate, and is also at Barrow Hill in the care of the same group. (Author)

Opposite page: On 13 March 1979, the driver of Class '25/2' 25130 was in communication with the signaller, seeking authorisation to move off Toton depot. The driver of 'Peak' 44008 *Penyghent* was climbing down from his cab, ready to conduct the same procedure for his allotted diagram. 'Sulzer' 25130 was built at BR Derby Works and introduced into service at Toton depot during July 1964, numbered D5280. Withdrawn from service in December 1982, it was cut up during March 1987 at Swindon Works, after spending over four years in store there. 'Peak' 44008, also built at BR Derby Works, emerged in December 1959, numbered D8, and was initially allocated to Camden depot. After periods of evaluation based at Camden, Crewe, Derby, Toton, and Nottingham depots, it settled down to working exclusively from Toton depot from May 1962 until its withdrawal from service during November 1980. It is now appropriately preserved on the Peak Railway near Matlock. (Author)

44 008
10

Above left: Several former BR and industrial locos were present at Booth's Rotherham scrapyard on 22 May 1989, and the Class '11' 12099 was certainly prominent in its orange livery. Built at BR Derby Works in March 1952, it was allocated to numerous LMR depots, including Nottingham and Willesden, before joining the NCB fleet in 1971, working in Wales and Yorkshire, prior to its despatch to Booth's yard from Bowers Row Disposal Point in West Yorkshire. It was in the company of another erstwhile NCB loco, former Ollerton colliery's Rolls-Royce 'Sentinel' 0-6-0 diesel-hydraulic *N.C.B. D8* (Works No.10257 built in 1966). 12099 was sold to the Severn Valley Railway in 1990, since seeing use as the Kidderminster yard pilot. The 'Sentinel' was not so fortunate, and being of non-BR pedigree did not attract such attention, and was broken up almost immediately. (Roy Burt)

Below left: Class '55' 'Deltics' 55016 *Gordon Highlander* and 55015 *Tulyar* (with the distinctive Finsbury Park TMD-applied white window surrounds and painted name) with 40019 *Caronia* at the end of the row, languish in Doncaster Works scrap sidings on 14 August 1983. Despite an apparent fate ahead of them, both were saved from the cutter's torch, and passed into preservation. However, the former WCML stalwart was not so fortunate, having been taken out of service in December 1981 at Springs Branch depot, Wigan, suffering with frame fractures, and was disposed of there during February 1984. (Roy Burt)

Opposite page: It is a tragedy that the original LYR steam MPD did not survive long enough to fall into the hands of the heritage ELR. What a splendid historical and useful depot it would have been today had it not been demolished. Classified as 26D and subsequently 9M from 1963, BR officially closed it on 12 April 1965, but it subsequently found use as a storage facility for 'mothballed' electric locos, including the veteran *Tommy*. This was the scene on 10 April 1971, with E3037 and E26000 *Tommy* on the extreme right of the line-up. For the record, the following Class '83' and '84' locos, and the sole Class '76' recorded on this visit, in the order in which they were noted down is as follows: E3036, E3040, E3041, E3039, E3045, E3028, E3037, E26000, E3027, E3044, E3035, E3098, E3025, E3024, E3029, E3031, E3030, E3026 and E3099. Despite its historical significance, progenitor Class '76' 26000 *Tommy* was cut up at Crewe Works in November 1972. The Class '83' and '84' AC electrics were eventually returned to traffic after works attention, the Class '84s' finally succumbing at the end of 1980, but most of the Class '83s' stored at Bury saw service until July 1983. (Author)

An overview of Barry's 'south yard' on 22 August 1970, viewed from atop the cab of Bulleid un-rebuilt 'Battle of Britain' Pacific 34072 *257 Squadron*, now preserved on the Spa Valley Railway. To the left can be seen today's West Coast Railways' main line stalwart, Stanier '8F' Class 2-8-0 48151, and immediately in front of 34072 is BR Standard '9F' Class 2-10-0 92207, which in 2022 was secured by the heritage Somerset & Dorset Railway as a long-term restoration project. What a significantly different heritage railway scene we would have today had it not been for the astonishing survival, foresight, and sale, thanks to Dai Woodham MBE, of so many locos from Barry, notwithstanding the subsequent painstaking restoration work undertaken, having returned so many to working order again, providing us all with much enjoyment for many years to come. (Roy Burt)